LISTEN
TO
THE
TREES

LISTEN TO THE TREES

Don MacCaskill

Drawings by Robert Tindall

Whittles Publishing

Published by
Whittles Publishing
Dunbeath
Caithness, KW6 6EY
Scotland UK

www.whittlespublishing.com

ISBN-10 1-870325-34-6
ISBN-13 978-1-870325-34-9

Page layout by Scotty-Dog Pictures

Printed and bound in Poland, EU
Produced by Polskabook

Dedication

This book is dedicated to all those who care for trees

Acknowledgements

To the Forestry Commission who, long ago, gave me a reason for living.

To my wife Bridget for transforming my roughly assembled words into the sensitively edited book you are about to read.

Poem

Glimpses in the Woods

A wet and rutted road winds through the woods
where glassy pools shine mistily alight.
The tender green of larch against the dark
and sombre hue of spruce is beauty rare
while tapering trees lead the enquiring eye
to hillside heights clad with their splendid growth.

The silence and the mystery of the woods
lie all around to hold the mind in awe:
what mystic spells, strange gods, forgotten rites
are held in that dark silence?

Except the whisper of the gentle rain,
no sound at all save where a tumbling burn
in haste to meet the loch goes racing down,
making a tawny mirror for the sky.

Rain dies away and, slanting through the clouds,
pale sunshine resting on a distant hill,
lighting a misty slope, captures the eye,
while flickering through the leaves, a watery gleam
touches the raindrops on each sodden bough
glistening and trembling as they slowly slide
into the wetness of the earth below.

Then, breathless moment! Framed by those proud trees
and quite as proud as they, an antlered stag,
caught unsuspecting in a sunlit glade,
gazed down an open pathway through the woods.
In calm indifference, serene and unafraid.

I.A. Cavenaugh (On her first ever visit to a Scottish forest in 1972)

CONTENTS

Foreword   xi

Introduction   xiii

Chapter One   A Village Set   1

Chapter Two   Road to a Forest   22

Chapter Three   Coillessan   40

Chapter Four   Old Woodsman   57

Chapter Five   Sheeps is Cheerier   76

Chapter Six   Fire in the Forest   96

Chapter Seven   Wild Work   115

Chapter Eight   Cattle Calamities   134

Chapter Nine   A Sheepish Exit   152

Chapter Ten   The Last of the Waterfalls   170

Epilogue   182

Afterword   186

foreword

DON MacCaskill was an outstanding naturalist, a brilliant photographer and a highly experienced forester. He was also a modest man, and it was no mean feat of his wife and co-naturalist, Bridget, to persuade Don to tell his story.

Don was always eager to inspire younger or less experienced naturalists and to help them to get to know the Scottish landscape that he knew and loved so well. *Listen to the Trees* is a beautiful account, strongly anecdotal and filled with humour, of a boyhood spent on the shores of a corner of Argyllshire, south of Oban, and continuing on into adult life in other parts of Argyll.

Don became a forester after fighting off illness, the setting-free of a trapped wild cat transforming his life and putting him on course to appreciating all the beauties, intricacies and stark realities of nature.

To Don, conifer trees were not just a cash crop – they were a place where all nature flourishes. He became a watcher and carer of wildlife, from the golden eagle to the badger, the raven to the ant, the red deer to the curlew, the oyster catcher and the peregrine falcon. He was an advocate of the woodland planting of mixed species so that the edges of conifer plantations and the islands within them are covered in birch, oak, alder and rowan to create beauty in diversity. Don saw the forest as a living world, and brings that world to his readers. Indeed, there are echoes of Henry Williamson's experiences in Don MacCaskill's writing – the same high level of observation, the same caring nature, the same sensitivities and the constant presence of 'soul'.

'Nature never betrays the hand that loves her' said Wordsworth. In his unique story, Don takes the reader inside the grass-roots commercial life of the Forestry Commission and the social make-up of forestry communities, weaving in their strategies, successes and failures. Don was not one of those naturalists who loves animals and yet is hostile to people. His relationships with other foresters and local characters are honestly and movingly chronicled.

He had adventures, too! Just look at the wild bulls. All in all, this is a very special book. Here is a knowledge of nature and a life of real success gently told by a man who contributed to making Scotland a better place. He had the 'seeing eye', as the Gaelic phrase puts it. This book will endure, and deservedly so.

Dr Rennie McOwan

Dr Rennie McOwan is a full-time writer and broadcaster, specialising in Scottish history and literature, mountaineering, and environmental and conservation topics. In 1996 he was given an honorary doctorate by Stirling University for his contribution to Scottish writing and culture. The following year he received the Golden Eagle Award from the Outdoor Writers Guild for access campaigning and his Scottish and outdoor writing, and in 1998 he was given a Stirling Council Provost's Civic Award.

introduction

I LOVE TREES. All kinds of trees. I can't help it. Look in awe upon some ancient oak perfect in shape and hundreds of years in the growing. Feel history in the Caledonian Scots pine, remnant of an older forest and witness to many a dark deed in a Highland past. Discover with delight a glade of silver birches, branches bending to a breeze, leaves quivering in the sunshine. Lay your hands upon an ancient stem, look up its gnarled length to a heavenly canopy above, and receive both inspiration and strength. And so on, through every wood and forest there is. Trees are surely the supreme example of a life-force stronger than our own. Some, like the giant redwoods of North America, live for thousands of years. Some, like our own oaks and pines, may live for centuries. All, given the right conditions, will regenerate their species and survive long into the future.

After many a twist of fortune, and a spot of luck as well, trees became my life. I trained to be a forester. In me was born a growing empathy with all things in nature – the trees and plants, the mammals and birds, the insects. I think I became a naturalist! I also became increasingly concerned for the damage man had done, and was doing, to the hills and glens of the Highlands of Scotland. Tree shelter, with the cover it provided for many a wild creature, had almost vanished. Over-grazing by sheep had seriously damaged much of the remaining vegetation. Erosion, caused by a wet climate, was adding to the devastation in many areas. The bare bones of a gracious and majestic landscape were becoming all too familiar and would, in a not too distant future, be part of a desert of rock, scree and peat. A wet desert.

This was a gradual realisation, of course, not the immediate con-clusion of an enthusiastic young forester just out of college. But in the years that followed I felt, in sympathy with many other environ-mentalists, that trees were the answer. There must be trees, forests and forests of them, woods and woods! They would hold the precious soil in place, become the very lungs of a polluted world, and eventually encourage other vegetation to grow. That became my philosophy; the splendid certainty of a young man, who, over many years, has not found it necessary greatly to change his mind.

My forestry was learned in a school where the true principles of silviculture were taught and put into practice. Individual species of tree were planted where they would do well and where they enhanced the surrounding landscape. The end result was pleasing forests of mixed species: Scots pine; European and Japanese larch; the three firs Douglas, *Abis noblis* and *Abis grandis*; Norway spruce and others. They satisfied the exacting standards of dedicated foresters and delighted country-loving members of the public who cared to wander in them.

This is the story of my first year as a young forester, the start of my career. I began my career in Ardgartan, a remote area which at the time was not easily reached by road and not greatly influenced by affairs outside the confines of its small village and huge forest. It was a community of true country folk, widely diverse in character and personal circumstance, self-reliant to an inventive degree and truly compassionate when a neighbour was in trouble. Nearly all of them, both men and women, worked in the forest. Among their number were expert woodsmen who often had a genuine interest in and concern for trees. Many would envisage a good result from their labour in a future time, long after they themselves had departed for a heavenly forest. They taught me much and it was one of the best years of my life.

Nowadays, Ardgartan is part of the Argyll Forest Park. A forest enterprise with a policy of landscaping in tune with the contours of the hillsides and planted with diverse species both conifers and deciduous. When passing through on my way up The Rest And Be Thankful, I look at it with a critical eye, and approve!

one: a village set

THE ancient van with creaking bodywork and protesting engine made slow and painful progress down the lochside. Its equally venerable driver wrestled manfully with the steering to find passage over the rough forest road. As we pursued our erratic course, sliding sickeningly, seemingly brakeless, into water-filled potholes and climbing with shuddering effort out again, he grumbled for the welfare of his broken-down steed and threatened to charge me for any damage it might incur.

'Where're you from?' he enquired eventually, in more friendly tones as we hit a smooth stretch.

'Kilmartin,' I replied. 'Know it?'

'Na. It's too far.'

As well it might be for this old wreck, I thought, for it would need to negotiate many a steep climb over the mountains on narrow roads with hairpin bends. As we trundled along I noted a wild and lonely landscape and decided this looked like the sort of place I had been hoping for. November sunshine glinted on the rippling waters of the long, narrow loch, a stiff breeze whipping them into little white horses which chased each other to the shore; they clawed greedily at the rocks and sent spray leaping high into the air. Woods of oak and birch were adjacent, all burnished bright in the golden glow. In startling contrast, and higher on the hillsides, were regiments of spruce and pine, evergreen and dark, broken up here and there by copses of larch, paler green turning to yellow. Above the forest, washed-out grasses, blackened heather and burnt-sienna bracken climbed higher still to sombre moorland. The mountain tops were snow-covered and pristine, unsullied as yet by man or beast.

'Strange place to come to,' continued my friend, unwinding a little with the need to chat. 'Naebody here and the village a long way back. It's lonely.'

'Suits me fine,' I replied.

When, at long last, we pulled up beside an old farmhouse at the end of the track, I could hardly believe my luck. Three miles from the metalled road and as many from the nearest habitation, it was the home I had been hoping for within a forest, lonely and alone. Built of old granite stone, it looked as though it had been there forever, battered by the elements but defying the worst they could do. Rippling wavelets washed the pebbled shore of a small bay, two rugged points crowned with pine and birch its sheltering arms. A wooded hillside behind rose steeply to moorland and mountain. Three old pine trees brooded over its ancient walls.

The old man helped unload the van with almost indecent haste, unceremoniously dumping my few odds and ends into the yard. Then, shaking his head sadly as if he feared for my sanity, roared away in a cloud of exhaust fumes. His departure did not bother me. I looked at my bits and pieces, all scattered around, and immediately decided to go exploring. Settling in could come later.

An old barn on the far side of the cobbled yard seemed a good start. I tried the door. It opened with a small protesting creak to a magic place, the old stables. They were dry and airy, the sun pouring welcoming light through slits in the old walls and as I shuffled slowly through a thick carpet of dead straw, dust rose in clouds to dance a dervish in sunshine shafts. Sadly missing were the sturdy Highland garrons who long ago would have stood in each stall, tossing their heads and whinnying a welcome. I could almost smell the warm scent of their breath and the aroma of golden hay stacked nearby. At the far end of the long building, ghostly skeletons of abandoned farm machinery were stacked in unlit obscurity. The roof seemed sound, the old oak beams fit for centuries more, the walls solid, and the stone flags of the floor, polished smooth with long use, still firmly set in their original mosaic. This place should be used, I thought.

Beyond the buildings, their walls forming the fourth side of a square, I wandered into a sad old kitchen garden partially walled and fenced, overgrown and neglected. A riot of weeds in an ordered rectangle suggested vegetables once growing there and the rest was a tangle of dead raspberry canes, brambles and currant bushes. In a sheltered corner two aged apple trees had autumn fruit still clinging to their branches – mice and birds had been at work on scattered windfalls. An old henhouse, with a broken, slatted roof stood in another corner and three deserted beehives were forlornly unused.

I knew from the Ordnance Survey Map that the wood beyond the garden was called Coillessan – the Wood of the Waterfalls. Better take a look. I crossed on stepping stones over a chuckling burn, climbed a steep bank studded with rocks and fern and found myself in a delightful wood all dappled in sunshine and shadow. It was mostly of oak at least two hundred years old, branches and stems gnarled and twisted, leaves, russet and golden, curled and crisp, still clinging to parent branches. A gentle breeze eased reluctant singletons away which fluttered slowly, reluctantly, to the ground. Ashes and birches, and rowans with burdens of bright berries, were dotted here and there. Alongside the burn alders, almost bare of their dark leaves, surely indicated wetter ground. Some distance away, seen through gaps in the trees, stark skeletons of oak were witness to a once more impressive wood. On the rising slope to my right, silver birch predominated and beyond it larch, planted many years ago, climbed the hillside towards a ridge. To complete this autumn ambience, the loch was ruffled gold in the sunshine and a young buzzard circled overhead, calling wistfully to parents who would no longer be feeding it. A raven pair 'cronked' to each other over a cliff.

But where were the waterfalls? I kept coming across small burns, each pebble-bedded, their waters bubbling merrily down to join the loch. Each must have found its way down that steep hillside and, presumably, over the falls I could not see. They must be concealed beneath the dense cover of trees. I turned up the side of the next little stream.

At first it was easy walking over springy, leaf-strewn ground, the only sound the murmur of hurrying water. I noted the stems of oak all encrusted with lichen and moss – obviously a wet place this and a humid west coast climate. But soon the burn's course became a gully, broad at first then narrowing to sides of giant boulders and splintered rock. Now its waters were splashing musically over waterfall steps to pools beneath. Swirling into ever expanding circles, they gathered momentum to tumble over and down to the next. Gradually the canopy closed, the brilliant sunshine was excluded and the gradient became much steeper. The gully was now a deeply shadowed place and a little sinister. I imagined the burn in spate, when sound would drown out thought, a monster who would sweep away all who might dare to venture in. Soon I was trying to find foothold on the steep sides of great water-polished rocks slippery with dripping moisture, lichen or moss. Often I had to deviate into the wood.

In about half an hour, I was suddenly out of the trees on a plateau-like area of peat bog. Here must be the sources of many of the burns I had encountered. Beyond, young larch marched onwards up the hillside to outcrops of rock on a heather moor. I turned along an edge of reeds and water-logged ground and soon found the next gully down the hill. As I stumbled down its precipitous course, I thought about Coillessan. Many burns and many waterfalls. Heavy rain would quickly transform each into unruly white water, racing and roaring to the loch. The wood was well named.

As I began to walk back to the house, bright light momentarily vanished and the sun was hidden by fleecy cloud. I checked the sky. To the south it was dark and threatening. Rain! None of my furniture or precious books were under cover. Waterfalls forgotten, I raced for home and reached the yard just as the first drops were falling. The nearest door, thrown hastily open, led straight into a large room. It was the kitchen. I dragged my stuff in, slammed out a downpour, then stood looking at what was to be my home.

Well, I had been warned. Two houses had been available, a modern, well-equipped one in the village four miles away and this farmhouse which had not been occupied for many years and needed a lot done to it.

A remote situation was what I had wanted. Here it was, a home old, neglected and sad and as dirty and dusty as you might expect. Spiders' webs decorated each corner and bluebottle corpses littered window sills and the floor beneath. The rafters above the kitchen range were blackened with the smoke of ages and I reckoned its hanging swivel could easily have held a whole roasting pig. A deal table in one corner, with a bench on either side, conjured up a homely farmhouse scene, the table laden with food and the cheerful family all seated around discussing the happenings of the day. I invented the busy, bustling life of a crofting household, the husband with his sheep on the hill, a cow or two in the byre, and the fishing to supplement his income. He had a placid, capable wife and three or four healthy children who all helped with the work on the croft. More likely, of course, it had been a tough existence, a struggle to make ends meet and unending drudgery. Never mind. The 'feel' of this room was good. It had been the focal point of all their lives and so it would be of mine. I even peopled it with a rabble of collies, a cat or two, and hens and ducks in the yard outside.

There were two large rooms on the ground floor. Each had an old Victorian fireplace with a cast iron grate and soot decorating the gaudily patterned tiles. A broken-down settee in one and two decrepit dining room chairs in the other completed the furnishings. The front door would not open and beside it a tiny cloakroom with a hole in the floor suggested dry rot. A stair with creaking boards and no handrail led upstairs. Two rooms matched those beneath. Each had a dormer window with a magnificent view over the loch, one an old brass bedstead with broken springs, the other a deserted wardrobe. Sandwiched between was a dark box room full of junk. To complete my palatial residence, a large bathroom boasted a marbled washbasin with brass taps, the oldest and rustiest of bath tubs, and a 'throne' of impressive size encased in warmly marked mahogany.

Downstairs again, I collected bits and pieces of wood from outside and after a few attempts got the kitchen range roaring. Essentials to unpack for today were cooking utensils, a plate to eat a meal from and, later, my camp bed on the kitchen floor. The Tilley lamp was lighted and in a little while the room warmed to a range now burning well. Soon I was enjoying a meal and thinking about the next day when I would report to the Head Forester and learn what work I was to do. The Kilmartin spoken of to my ancient driver of the morning seemed, at that moment, to be a long way away and a long time ago.

The Kilmartin experience, however, was not that easily packaged and put away, for it explained the present. The route to Ardgartan had been tortuous and long and forestry not my original goal. But fate had taken a hand and here I was, ready to put down roots. Already it seemed that the infinite wonder and silence of the mountains, moors and woodlands around Kilmartin, together with their wild inhabitants, were to be discovered here. That was good, for they were an integral part of the creature that was me and would always remain so.

The village where I grew up, set in craggy woodland, was made up of twenty-five old, stone-built houses and six, incongruous, red brick 'semis', all in a rough semi-circle alongside 'Front' Street and the Bruaich. The Bruaich was the village green. The church, built in 1835, the second on the site of a much older one, boasted two famous ancient crosses and an unusual, possibly unique, archway of stones on which were inscribed the names of local men fallen during the First World War. The church, together with its manse, dominated the scene and was a reminder of our mortality. The Hotel, almost opposite, was an old coaching inn, rambling, stone-built and painted white, the centre to which migrated all those in need of worldly consolation. They came, especially on Friday evenings and Saturdays, on bicycles or in old bangers, to celebrate the end of the working week. A crocodile of folk, the minister among them, all wending their way from church to hotel, would mark the occasion of a baptism, wedding or funeral already taken place and about to be suitably commemorated. If it was a wedding, the dancing and merrymaking would go on all night.

The ground fell steeply away from the village, south and west, to farmland, rocky outcrops, wooded knolls and desolate marshland. Beyond that horizon, a boy could imagine mysterious and lonely islands in a vast unending sea. To the north and east, wooded hills climbed, step by step, to rock-crowned mountains often veiled in dark curtains of rain. These were the 'Highlands' to the kids in the village and only attainable by foot or bicycle power. The River Add rose in the hills to the east and meandered, with many a long change of direction, into Loch Crinan four miles to the south. A tributary, just below the village, angry in flood or drifting sluggishly in drought, was the Kilmartin Burn, a great place for the fishing.

And history was there, too, a powerful influence which made us all feel we sprang from a romantic, often violent past and that somehow we were still a part of that past in the present. The village had once been the site of an early settlement in the ancient kingdom of Dalriada. Its crumbling chapel, associated with St Columba, the old ruined castle close by, the weatherworn duns, standing stones, cairns, cists and cup-and-ring markings dotted around the area, all were monuments to the early settlers who were our forebears. Brooding over all was Dunadd, a rugged rock fortress standing stark and lonely on the marshy wetland to the south. It had guarded the River Add from invaders and was once the capital of the Kingdom of the Scots where their kings were crowned. Little of the fort now remained but the old stone basin, in which each was purified before the ceremony, was still there. A beautiful carving of a wild boar, painstakingly sculpted into a rock nearby, fascinated me and prompted thoughts of long-ago strange wild beasts in the woods surrounding my home.

The village was part of a large estate and its laird was monarch of all, almost literally ruling over our lives. Though he spent most of his time in his London residence, his factor managed the estate and controlled the leasing of the many farms and other properties on it. A humane and kindly man, he was inevitably hamstrung by the wishes of his employer. Woe betide any tenant who fell out of line or was caught poaching, for the consequence could be the loss of his job and eviction from his home.

The laird and his family came home for the shooting, the stalking and the fishing. Large numbers of guests would arrive at the House and the whole village, to a greater or lesser extent, was then in turmoil, the quiet routine of our lives upset. Our mothers were employed to bake, prepare gargantuan meals and to attend to the making of beds and the general cleaning. Eleven gardeners worked all hours in the huge kitchen gardens to produce the fruit, vegetables and flowers that would be needed.

Salmon, venison, grouse or pheasants, whatever was in season and often the trophies of the guests, were all brought in by the keepers to be cleaned, hung if necessary, and, in due course, prepared for eating. Home-reared poultry supplemented the game and local farms also provided milk, butter, cheese and eggs. About the only amenities missing were a vineyard and a distillery!

Christmas and Hogmanay were particularly busy times when the House would be full of family and friends. The laird gave a party in the Village Hall, a fine building which he had gifted to the villagers but later sold when hard-up for cash. It was enjoyed by all, for there was always plenty of food, a good band and much energetic dancing, and presents for everyone. All the guests were expected to attend and our laird always arrived in full Highland dress, a silent, unsmiling presence in full regalia, who would watch the festivities for a short while then, duty done, take his leave.

My parents ran the Post Office and there were two shops, Ross's and Ramsey's. Each was an important focal point of village life where you could get up to date on gossip and learn what was going on where, when and, especially, with whom! There was a piece of history there, too, for my grandparents had owned the Post Office business, grandmother to run the office, grandfather to collect the mail from the steamer at Ardrishaig and take it, by road, to Oban – a 40-mile journey in a four-in-hand with a change of horses and driver at Craignish, the halfway point. By the time my parents took over, the mailbags were carried by the local bus from Lochgilphead, collected by our own two postmen and brought into the office to be sorted – much gossip and grumbling if the weather was foul. The job done, off they would set on their sturdy Post Office bikes to cover miles of rough roads and tracks. Letters, small parcels, and often, out of the kindness of their hearts, 'messages', small items ordered from the shops, were all delivered to a scattered community, the job sometimes taking all day.

The postmen, in fact, were the thread which linked events and gossip in the village to the farms and steadings to the north and south of it. Old MacMaster lived in Front Street. He had been a sergeant in the Army, had been badly wounded and demobilised, and lived almost entirely on whisky. He had no wife, spent all his wages on drink, sold all his furniture to that end, and had to make do with a covering of mailbags on the floor for his bed. In spite of this, he did his job well enough for the most part, delivering all the mail north of the village. Dugie, ginger moustached, bright and full of fun took care of mail for the south. And when Old MacMaster finally succumbed to his inevitable ending, he was replaced by a man who a was total delight to all the lassies in the village. Lachlan was young and handsome, his thick crisp hair always in place and his uniform immaculate. He was single. One day, bicycling over the Kilmartin Burn in a gale, the wind filled his vast Post Office cape and blew him, cycle, mail and all, into the raging waters. My father fished him out, just in time. He recovered from this alarming experience and from a nasty chill but, as a result, became even more desirable – a young man who obviously needed a wife!

The Post Office was the family home as well. It was cramped, with no privacy, but comfortable enough to a youngster who knew no better. Access to the small office was from a side door in the front porch. A large room behind it, filled with cubby holes and tables, was the sorting office. The kitchen, with a range on which all the family cooking was done, its fire never out, was our living room. It was snug in winter but often too warm in summer.

Behind was a small bedroom, that of my parent's; my father, a rather quiet man who seldom spoke but read a great deal and my mother, who was absurdly generous and passionately caring of anybody in trouble, especially Old MacMaster down the road. A tiny box room was my sister Peggy's. My younger brother, Peter, and I slept in the sorting office because there were no more bedrooms. There was no running water. This was obtained from a tap in the yard, so there was no lavatory or bathroom but a dry closet outside and kettles of water perpetually boiling up on the range.

It was a small world, too, that we lived in, narrow, parochial and in the shadow of the Estate. The little affairs of each and every one of us could loom unhealthily large and inspire intense curiosity, the reverse of the coin being real concern and helpfulness in adversity. There was little contact with the outside world. Villages, even within a few miles, were seldom visited and a trip to Oban in the rickety old bus was a great day out.

Social occasions were neighbourly 'droppings-in' for a cup of tea or a dram, and a gossip, a ceilidh. Football matches were played on the green and afterwards the players and many of the spectators repaired to the hotel where Peter McLennan, the proprietor, presided over the bar and often consumed rather more of the product he was selling than he ought to have done. That is, if he could evade the eagle eye of his missus.

The village bobbie lived next door to us and was a presence my mother threatened when family discipline got out of hand. Angus was a stout, burly man who kept himself very much to himself and had a reputation for being strict. His job was mostly a light one, however, for there was little or no crime. At relevant times of the year he would cycle many miles each day to make sure that the local farmers were dipping their sheep and were keeping their cattle clean. Otherwise his most onerous duties were on Fridays and Saturdays when all the world converged on the hotel. At 9.00 p.m. sharp he would take up a position in the road outside and anyone not out on time was in trouble. Our bikes were an obsession for him. They must be looked after properly and be in good nick and a stern reprimand was given to anyone who dared carry a passenger on the crossbar. He caught me at it, one day, and was so concerned that he reported the matter to the Inspector at Lochgilphead. To my mother's dismay, that august gentlemen turned up to give me a lecture.

To us youngsters, with not a great deal in the way of entertainment available, a burial was a great occasion! The graveyard was not large, the soil not deep, and finding space for everyone, in an age when cremation could only take place in distant cities, required meticulous planning. It was the custom to 'double-deck' the coffins. So, when Old George the beadle arrived, the tools of his trade balanced on one shoulder, his faithful old terrier alongside, we all trooped through the wrought-iron gate full of anticipation.

'Is it a top layer job today, George?' we would ask.

'What's it to you?' he would growl crossly.

With a suitably solemn air to fit the occasion, and jabbering away to his dog in Gaelic, the old man laid out his tools, one by one, on the ground. Ignoring his delighted audience, he then spat on his hands and got down to work. Our moment had come. Each spadeful of soil was methodically selected, to create the final shape of the grave, then lifted out and carefully laid aside for replacement. Each was checked over, as a gardener might look for weeds. Rotting pieces of old coffin, mildewed brass plates and handles, stones, all were fastidiously picked out and tossed nonchalantly over the wall. They landed on a pile of unwanted debris which grew with each funeral. More ghoulish remains were there, too. Bones. These our worthy beadle sifted out, knocked on a stone to remove the last of the soil, then sent flying to join the rest.

Afterwards, we examined each offering with interest, learned quite a bit about the human skeleton and noted that once dead our species was of no great importance.

It was in Kilmartin that my interest in wildlife was kindled, though I never heard the words 'conservation' or 'wildlife' as such. The former, if it happened at all, was only concerned with the safety of the pheasants or the grouse which guests on the Estate would eventually shoot. 'Wildlife' was the small birds whose nests we raided each spring and summer and the fish which we tried to catch in the two burns. The matings of wild or domestic creatures, mammals or birds, were observed without wonder or thought for the young that would be born, or how they would survive. Carcasses, come across in the woods and fields, aroused no particular curiosity as to the cause of death – they were briefly noted then left to rot or be consumed by scavengers. 'Vermin' were the foxes, wild cats, badgers, otters, stoats and weasels, who would be poisoned, snared, trapped and hunted, because of their supposed predation upon domestic animals, or the animals and birds the Estate required for sporting purposes. But at last, and perhaps because of an event one night when the whole family was roused, I did begin to wonder and to question.

We were woken by a dreadful caterwauling. It came from the nearby wood and was so horrible it sent a shiver down my spine. My father came running to collect his boots.

'Want to come, boys?' he threw at us.

'What is it, Dad?' I asked.

'Don't know. Hurry up.'

The whole family set out, my mother bringing up the rear because she would not be left alone. It was the blackest of nights and spooky. The paraffin lamp swept inadequate light from side to side and cast long weird shadows which were never still. Intermittent screaming from the wood painted fearful pictures in my mind of a creature in agony. My father walked quickly and we scrambled to keep up. First, the Bruaich – easy because we knew every inch of it. Then a slithering, sliding progress through thick bracken, down to the burn – slimy stepping stones, splashing, and water in my 'welly' boots. Finally, a scramble through rough rock, fern and scree which brought us up a long bank to the edge of the wood. There, we all paused to listen.

No yowling, now, the animal quiet, perhaps hearing our approach. Or, dead. I looked about me in dismay. How could we ever find it? Sentinel tree trunks, dimly seen, were ghostly guardians of impenetrable obscurity. Birch leaves, rustling softly in a silent world, fluttered delicately above our heads in the flickering light but could give us no clue. Which way to go? All at once another shriek echoed through the wood. I shivered. Our quarry was still alive, but in terrible trouble.

'It's hurting,' I cried in distress.

'Be quiet,' snapped my father and set out along a well-trodden path through the trees.

'It must be a rabbit,' I whispered, knowing well how they could squeal.

'That's no rabbit,' he replied. 'Come on.'

Suddenly, low growling replaced screams of rage. We crept slowly towards it. My parents signalled us to halt. The growling ceased and I held my breath. The lamp steadied, moved to right then left, returned to pinpoint a spot, and remained fixed. At last.

'It's a wild one,' Dad exclaimed.

An enormous cat was caught in a gin trap. In the remorseless beam of our lamp, with ears flattened and lips drawn back, it crouched low to the ground growling fiercely. Held by the brutal gin, it tried to draw back but could not. When my father made cautiously towards it, growling exploded into spitting hate. Riveted, I took in green eyes reflecting fear, dark stripes on a fawn brindled body, and bristling whiskers drawn back in a ghastly grimace. Dad bent closer.

'Don't think the paw's broken,' he announced. 'Just torn.'

'Do be careful,' said my ever anxious mother.

It all happened in a flash. My father threw his jacket over the animal and released the trap's spring with his boot. With an awkward movement, half leap, half struggle, the cat sprang clear. As it limped off into the kindly cover of the night, I noted the broad, dark rings on the short, thick tail. Then it was gone.

'That would be one of Jimmie's traps,' remarked Dad, thoughtfully.

Jimmie, the local rabbit-catcher, made quite a good living out of the rabbits which inhabited our hillsides in thousands. He snared or gin-trapped them and, like everyone else, evidently gave not a thought to the cruelty involved. Some were sold to our local shop, but most went to the butchers in Lochgilphead, our nearest large town. As we tramped home in silence, I remembered the sheer beauty and grace of the wild cat, even in its predicament, and thought it could have been a long time dying. What harm had it done? What other creatures, beside rabbits, did Jimmie catch? And so on. It was only a fleeting questioning and the answers, had I known them, would not have much bothered me then. That was for later. For the present, it was enough that tonight there had been a happy ending.

Gradually, however, my interest in wildlife broadened beyond the collecting of birds' eggs. Small fish caught in the Wee Burn led to thoughts of the magnificent salmon which swam up the Big Burn to spawn. In shallow pools we practised sliding gentle fingers beneath their bellies until, lulled into stillness, they could be lifted quickly to the bank. The cock fish were easily recognised by the hook on their lower jaws but care was needed with the hens in case the movement of our hands stimulated premature ejaculation of their eggs. These fish we always put back in the water. One day, Harry-the-Englishman showed me a large salmon whose neck had been bitten and some of the flesh torn away by very efficient teeth. A fox, I suggested. But, no. He had found it on the river bank and right beside was the 'spraint', or dropping, of an otter.

That was the culprit. As a youngster, therefore, I grew up to think of these delightful animals as villains who preyed upon the precious salmon, raided chicken runs and were even supposed to take lambs. They were to be shot, if possible, or trapped for their valuable pelts.

During the spring and summer we woke to the sound of the corncrakes. Each day the air was filled with their strange ratchet song in the hay meadows. In a way, it was background sound, like bees humming or grasshoppers clicking, to which you paid no particular attention and in time hardly heard. One day, when I was standing on the bridge over the burn watching the minnows below, their song was obtrusive and loud. Archie, the farmer, happened across. He was a tall, quiet man who took little part in the social life of the village.

'Ever seen one, Archie?' I asked, curiosity overcoming shyness.

'Seen what?' he enquired, apparently oblivious to the chorus.

'A corncrake.'

'You'd have a job.' He laughed and let himself through the gate into his field.

He was right. I spent a long time that day trying to spot those shy and elusive birds. The wretches threaded busily and unseen through the long grasses, never still for more than a second and all the time uttering their maddening song. No luck that day, and over the years, I spent many hours trying unsuccessfully to find a nest. The best achieved was fleeting glimpses of a streaked brown bird with chestnut on its wings, scurrying into willow scrub at the side of the field. Their numbers have declined over the years to near extinction, at first because farmers were careless at hay-making and they were killed and then because of modern agricultural practices which involved the making of silage before the young were out of the nest.

Haymaking was another social occasion and good fun. Of particular interest were the horses who were used at that time for most of the jobs involved. Archie had two Clydesdale mares called Jess and Mary, the first a beautiful chestnut the other a grey, so dark she was almost black, but with four white 'socks' and a star on her forehead. They were docile creatures who plodded patiently up and down the big field as the rows were cut, scarcely needing guiding so well did they know the routine. At the lunch break, as we all gossiped and ate our 'pieces', I would watch them contentedly munching oats from their food bags, thick coats glossy in the sunshine, wise eyes calmly regarding us mortals sitting close by. Once cut, the hay had to be tumbled so that the air could pass freely through to dry it. The great pair would set off again, this time with the 'tumbler' behind them, up and down, up and down the long rows. To them, it was all a part of the day's work and another bag of oats would be ready when they finished.

But the day's work might not yet be over. If the weather was fine and the tumbled hay already dry enough, they would be led off to gather it all into huge piles at the tops of the rows. To and fro they plodded once more, the farmer behind to guide them, heads down, steady as you go, tails swishing to defeat the flies. Now, us kids could play a part. The large piles must be built into stacks, about ten feet high and often on tripods which were three poles stuck into the ground and drawn together at the top. With bare feet and tough tingling soles we trod each bundle as it was added, tramping the hay round and round until at last, too high and tramping impossible, the final forkfuls were secured by ropes thrown over the top and tied to stakes on the ground below. In due course, if the weather was still good and the hay was truly dry, it was transferred to the 'slipe', a wooden platform balanced on two wheels. Then, load by load, often swaying perilously, it was drawn by Jess and Mary to the farmyard. There the sweet-scented winter food for cattle beasts was built into one enormous stack.

The horses were more friends than beasts of burden and I spent much time talking to them, smoothing my hands over their coats, and even, as a great concession from the farmer, learning to work them. Events in the lives of our equine friends were almost as important as those of our human neighbours and death, illness, a visit from the vet, the birth of a foal, all were noted, found interesting, and discussed. There was no surprise, therefore, when on the last day of school one summer term, a pal of mine whispered magic words as we filed into class: the stallion's coming tomorrow! When the local mares were 'in season', the stallion was brought to the village to serve them. The momentous event took place in the courtyard of the old hotel and Peter

McLennan always supervised, for the Clydesdale breed of horse was his great hobby. Unfortunately, he also considered these occasions most unsuitable for us youngsters to witness. The heavy oak doors of the archway to the yard were firmly closed and though knotholes in the old wood allowed a peep, that meant only tantalising glimpses of a dark body rearing and the sound of thundering hooves as the vital moment grew near. To obtain a proper view was a long-held ambition.

'Let's hide in the yard,' I had suggested. 'We can climb up to one of the windows in the barn.'

Alan's eyes danced but he was dubious. 'They'll find us.'

'Not if we're there in plenty time.'

The barn was one of several buildings which surrounded the cobbled yard of the old coaching inn. The archway through which all must pass to gain access was a part of the main building. As we slipped unobtrusively through the next morning, its resident swallows and martins darted backwards and forwards, swift arrows of feathered light, screaming shrill territorial messages to each other and taking food to their young. Suitable background sound for the excitement that was to come. A superb mare with gleaming coat, white socks cleaned and polished, and a white star on her forehead was standing there. It was Mary. Tied from her halter to a ring on the wall, she whinnied a soft welcome to old friends then shuffled her feet and regarded us with gentle, incurious eyes. The regular clip-clop of heavy hooves advancing up the street warned us that her mate was arriving. We slipped past her impressive bulk and in a few scrambling moments, with prickly, fragrant hay in our faces and our sneezes from the dust agonisingly suppressed, we were safely in position at a window in the barn.

Just in time. The stallion led by his owner, Donny McLeod, was already plodding through the archway. Mary whinnied and tossed her head. Peter and Archie came quickly from the bar where, no doubt, they had been discussing Clydesdale pedigrees over a pint. The proprietor, after a quick look down the street to make sure no kids were lurking close by, firmly closed and locked the great oak doors. We held our breath and struggled with itchy noses that threatened to give us away. Delicious shivers were chasing each other down my spine.

The stallion was becoming restive. He stamped his enormous feet impatiently, tossed his magnificent head and with flared nostrils whinnied repeatedly to the mare. He certainly knew why he was there! Mary became more and more restless, and though Archie kept soothing her and patting her elegant neck, she would not be still. The tension grew. Both animals were held with difficulty. Surely, the vital moment must be at hand. It was. The mare was suddenly quiet. The stallion, hind feet stamping, forelegs wildly thrashing, heaved himself up to cover her. There was stunned incredulity in two innocents when Peter guided the stallion's penis into the appropriate place! In urgent, thrusting moments, it was all over. The stallion slid his huge bulk away and Mary, free of his enormous weight, gave herself a shake as if to say: thank goodness that's over. She peed copiously, luxuriously, all over the smooth cobbles!

'That's a good sign,' we heard Peter remark.

A sign of what we had no idea, but the men disappeared into the pub, no doubt for a celebration, and we, pop-eyed and speechless, slid down from our eyrie.Our two Clydesdale friends were now standing quietly side-by-side, secured by their halters. They munched contentedly from the bundle of hay at their feet as if nothing unusual had happened.

'Let's go to Peter's shed,' suggested my friend, confident now we had not been discovered.

I knew what he meant. The proprietor was often to be seen slinking out of the hotel and into the yard with a bottle hidden under his jacket. This he would add to his secret store of screwtops, ready for when a great thirst came upon him. Nobody was about so we slunk along the side of the barn, lifted the catch on the door as quietly as possible, and slipped in. We each lifted a bottle from the shelf, unscrewed the top, emptied the precious beer on to the floor in a dark corner, then refilled the bottle with pee. We had done what we had come to do and in due course, Peter must have made an unfortunate discovery and may even have had some idea who the culprits were, but once safely home again, we never heard a word about it.

One day, my mother sent me to Ross's shop to pick up the messages she had ordered. The store was always busy for it sold just about everything anyone could possibly want. Groceries of all kinds could be bought at three of its counters, each watched over by a female assistant. There was haberdashery and clothing and even a resident tailor who ran up stout tweed suits to order. Household goods and all manner of ironmongery occupied another section, and wellington boots and tackety boots were placed on a rickety stand near the door. Eight people in all were employed. Nobody ever paid for their purchases at the time, but settled the bill at the end of the week. Sometimes. While I was there, Alan, my friend and son of Ross's storeman, came bursting through the door.

'There's a buzzard making a hell of a noise over the top of the Bull Rock,' he announced. 'Want to come and see?'

'What kind of a noise?' I enquired, only mildly interested, for the wistful calling of buzzards, 'pee-ou, pee-ou', could often be heard around the village; kestrels and sparrowhawks were also common.

'Sort of screaming,' he replied. 'It keeps diving at the ravens.'

Groceries forgotten, I ran for my bike. The Bull Rock was the local name for a crag, quite close to the road, in a nearby glen. The lower slopes were covered in a scattering of old oaks on a gentle rise. Then giant boulders, fallen from the heights above, made a wilderness of cracked and creviced rock amongst which ferns, nettles, dockens, and some wild flowers all grew. I knew of a fox den hidden in there. Then the cliff itself. It rose in dramatic, smooth and unclimbable rock to the heather-covered hillside above. Its face was broken only by narrow ledges, some bare, some with a dusting of plants or grass. Now, it smouldered in the hot sun of midday.

We had come quickly and Alan's bird was still there loudly scolding from somewhere high above the crags. That's no buzzard I decided as against a background of clear, blue sky I tried to spot it. Then, to my unaccustomed ears, it seemed there might be two birds calling, the strident protest coming, each time, from a different part of the cliff. Or, was that bird moving so fast it seemed like two? Then, I picked up another sound, this one instantly recognised. The ravens that Alan had mentioned were calling again, still making objection to something they did not like. It was maddening, though. I could not see any birds at all against the rocky heights nor in the wide expanse of cloudless sky above.

At last I discovered the raven pair, glossy black in the sunshine and flying close together across the cliff. They were quite near the top and seemed really bothered about something. Almost at once, two strange birds came swooping towards them, falling like stones from the sky and screeching loudly all the while. I held my breath. Surely, the ravens would be killed. But, no, there was no kill, no contact at all. At what seemed like the very last second, the attackers veered away with insolent ease, soaring up into the vast blueness above and then out of sight over the ridge.

The action had broken the ravens apart but they quickly came together once more and continued flying backwards and forwards across the cliff face. Only a small distance apart, they were 'speaking' to each other with deep 'disgruntled' croaking. Was there to be another attack?

There was. Suddenly, from somewhere out of the sky, the two streamlined birds came shrieking and stooping again. Faster than a buzzard dropping to a mouse, they tore towards the ravens with astonishing speed. I held my breath but, yet again, they broke away at the very last moment.

The strange birds kept on coming, attacking, harassing, shrieking hate. But the ravens were not to be driven away. They dodged and jinked from side to side, sank like stones to avoid the upward swooping from below and lifted easily, seemingly light as air, to evade each meteoric dive. All the while, they kept up their angry scolding and the strangers their raucous screaming.

But, all of a sudden, there was silence. No unearthly screeching. No croaking from indignant ravens. No action at all. The strange birds had vanished as if they had never been and the ravens had disappeared, too. It was eerie.

'Did you see that?' I turned to Alan, almost speechless.

He nodded, excitedly. 'Where have they gone?'

'Away over the top, I think. Let's wait a little.'

The ravens soon resumed their patrolling flight, gliding back and forth across the cliff. After a short time, they alighted on a ledge quite high up. At first I thought they were just resting but then I noticed a dark bundle of sticks behind the birds, tucked away beneath an overhang of rock. A nest! Their nest! Perhaps they already had chicks, though I could not see at that distance. Maybe the strange birds were nesting on the cliff, as well, and that was why there was some kind of competition between the two.

'I see them,' exclaimed Alan, suddenly. 'Go left from the ravens along the cliff and a bit higher. They're on another ledge.'

I picked them out, eventually, one bird on a bare rock shelf with a small overhang to shelter it, the other perched on a pinnacle of rock close by, preening its breast. No binoculars as yet, I decided the one on the ledge had its head down, maybe to chicks. Against the cliff face, the sun highlighted the dark, probably grey head and back of the first and the pale speckled breast of the other and when the one on the ledge lifted its head to look towards its mate, I could swear it had a hooked bill. Not a buzzard, though, nor yet a kestrel. Then the break-through. I re-membered Dugie-the-post recently reporting a peregrine falcon stooping high above his head as he cycled through this glen. I had thought he was pulling my leg, but that, surely, was our bird. All the way home, I flew with the falcons, swooping and soaring above the crags, monarch of all I surveyed beneath. This was the beginning of a lifelong interest in the bird.

Then, a final significant happening. One early morning, when we were kicking a ball about on the Bruaich before school, Jimmie Mac-Donald came by pushing his bike and, as usual, it was hung with the rabbits he had caught on the hill. This time he had a large sack on his back as well.

'Want to see what I've got, boys?' he called, evidently pleased with himself.

Curious, we gathered round thinking he might have caught a rare black rabbit or even a hare. With great ceremony, and some effort, he lifted the bag to the ground then snipped the string at the top. Pausing a second, no doubt to build the moment, he up-ended it. No rabbit. No hare. A large grey animal fell to the ground with a thump. It had dark brindled fur and white-on-black stripes on its long, slender head. I recognised it at once from a book I had at home.

'What d'you think that is?' Jimmie enquired, noting with satisfaction the stunned silence about him.

It was a badger ... once!' I replied, surprised at the revulsion I felt. 'How did you catch it?'

In the trap,' the rabbit catcher replied, matter-of-factly. 'Thought you'd like to see it.'

I looked at the bloodied head – our friend had bludgeoned the animal to death with his spade – the torn and broken foreleg by which it had been caught, the haunches still sleek and unmarked, the silly little tail. And felt sick. Why did human beings have to kill beautiful creatures like this? What harm did they do? Suddenly, I needed to find out.

two: road to a forest

FAR from Kilmartin, in Ardgartan, it was a restless night when my home responded to the new life within it with strange rustlings and creakings. After a quick breakfast, I rode on my bike the three miles back along the track that had so enraged my friend of yesterday. It was a lovely day, the glen as enchanting as before, autumn colours blazing, the loch sparkling and at peace and a light-hearted peddling to my first day at work. At long last, I was where I wanted to be and about to embark on the career I had set my heart on.

First stop would be the Forest Office. There I would report to Jimmie Reid, the Head Forester, and meet the rest of the staff. I reached a sign, ARDGARTAN FOREST OFFICE, at the entrance to a cobbled yard, and turned in. All was unexpectedly quiet. A large timber building to one side was presumably the office, but there were no men about, no vehicles, no ponies and nobody organising the day's work. I tried the door but found it locked. I looked through the windows. In one room there was a fine new desk, a tall filing cabinet and a couple of padded chairs – no sign, though, of the books, files, maps, papers, notices or general clutter one would expect to see. The only other was bare of furniture. I looked around, wondering what to do, and noticed a trim little house standing alone at the furthest corner of the yard. It was sheltered by tall larches, looked old, and had two beautiful, neatly pruned umbrella pines on either side of the door. A collie dog rose from a mat in the porch, yawned mightily, stretched himself fore and aft, then sauntered towards me wagging his tail. He inspected my legs, sniffing them carefully from knee to ankle, found them uninteresting, then wandered back again. I decided to see if anyone was at home.

There was no doorbell, but two knocks and seemingly endless minutes later I heard footsteps. The door was opened by a man of about fifty with a most unusual appearance. His face was deathly pale. He had no eyebrows or eyelashes and his ashen hue was accentuated by the unlikely red of an unruly mop of hair. That hair! From a curiously artificial 'line' across the forehead and a parting in the centre, it fell copper-coloured, long and lank, to well below his ears. A wig! Could this odd apparition be my new boss? The tweed jacket, plus-fours, polo-necked sweater, thick long socks and heavy boots suggested it was.

The man stood silent, a sardonic gleam in his eye as if he was used to the reactions of strangers. I stammered an introduction, but without a word he turned into the house indicating with a jerk of his head that I should follow. We entered a small hallway with a mirror on a stand for coats and a huge aspidistra reaching to the ceiling. A shadowy figure beside an open door smiled shyly, did not speak and melted away into the room behind, which I took to be the kitchen.

'The wife,' growled this extraordinary man.

I followed him into a dingy little room and he pointed to a chair on the other side of a decrepit table piled high with papers, files and books. As I sat down, a fleeting glance took in the ancient cabinet, with a broken leg, propped against a badly stained wall, a bookcase leaning tiredly against it, its shelves sagging beneath a miscellany of files, maps and a pair of binoculars, and an old paraffin stove which was belching forth evil fumes but having little effect on the cold and damp.

'Morning,' growled Jimmie Reid, at last. 'You must be Don MacCaskill. Excuse me a minute.'

And with this he disappeared once more. All ears, and by now intrigued, I heard the clink of a bottle against glass, imagined its contents thrown in a oner down his throat, and then the subdued squeak of a cupboard door being surreptitiously closed. He came back apparently now ready to face the day.

'Have you settled in all right?'

'Thank you, yes,' I replied, eagerly. 'This is a great place.'

'It's an old house,' he continued, ignoring my enthusiastic comment. 'I've arranged for a joiner and a plumber. Let me have a list of what needs to be done.'

'Thank you,' I said again, and awaited my instructions. There was a long pause during which my boss shuffled papers about and appeared to have nothing to say.

'What have you for me today?' I blurted, at last, embarrassed by the silence and anxious to get to work.

'Fact is,' he spoke, it seemed almost apologetically. 'I don't need anyone until the spring and then not someone just out of college.' He must have seen my face, for he continued hurriedly. 'Now you're here, I'd better find something for you to do. Get settled into your house. Take a look at the forest.'

'Is that all?' I asked in astonishment when he said no more.

'There's Norway spruce and a stand of Douglas to be thinned when I can spare some men. I'll put you in charge. The squad's over on Coinnach just now, fencing. A new planting. Meantime, there's Old John on the croft at Mark.'

'Old John?' I stammered.

'That's how he's known. Mark is down the loch beyond Coillessan. He's there to watch out for fire and is supposed to weed and keep the drains free. The area was planted two years ago. The old rascal lets his sheep in for a bit of free feeding. Keep an eye on him, will you?'

With that he picked up some papers and became deeply engrossed. Shattered, I could think of nothing to say and when the silence had become too long and no words of dismissal were uttered, I rose to leave muttering something about reporting again next morning. Jimmie only grunted. I cycled home, angry and wondering what had gone wrong. Twenty-seven years of age, qualified for the job and eager for the responsibilities it held, it seemed I was only fit to watch over a tiresome old man who would not control his sheep.

The path to Ardgartan had, indeed, been tortuous and long, but I felt I deserved better than this. Education has always been taken seriously in the Highlands and many of its sons and daughters progress to posts in government, education, the Church or industry. Not just a few rise to the top of their professions. So my father had dreamed for me. Good schools, primary and secondary, then university and a degree, would mean a career and a good job.

The beginning, of course, had been the Primary School in Kilmartin. Typical of its time it was a solid Victorian building, a teacher's house with a classroom on either side of it. You could not call our classrooms inspiring. Furnished only with desks, tables and a blackboard, each was bare of decoration. There were relatively few text books and no exercise books – we used slates and a blackboard. A roaring fire in an old iron stove, one in each room, was cheerful in the winter term but though these kept us warm they also created air that was humid and heavy from steaming wet jackets, socks and shoes. The toilets and playground were in the yard.

Nor was the teaching likely to arouse in us a burning desire to learn more. Mrs. Robertson, a kind, motherly woman, took the first and second year pupils. She listened to solemnly chanted alphabets and rhymes, and later taught us to read and write, to add up and subtract, and, girls and boys alike, to knit and to bake simple cakes and biscuits – the latter popular in a nice cosy kitchen where titbits were to be won. Her husband taught the seniors. He was hated and feared, for he had a violent temper. For something as simple as the wrong answer to a sum, we were shouted at until all rational thought had fled and quite often strapped as well. We learned the whereabouts and names of continents, countries and capital cities, with a few mountain ranges and rivers for good measure, the dates of the crowning of Scottish kings and famous battles fought against the English, English grammar – a total mystery to kids brought up in the Highlands – and literature, classic Scottish only. A good enough grounding, at that time, to see us all into secondary school. Unfortunately, it was thought that English was the only language in which you could get on in life, so the Gaelic, which most of our parents spoke fluently, would not be passed on to another generation.

School was not the place, either, where we learned about nature. Though manifestly all around us, it was not a subject considered important – in fact, it was not a subject at all. One day, when we were all shedding jackets and boots in the cloakroom, a chaffinch, hotly pursued by a large brown bird, flew through the open door. The chaffinch, jinking cheekily, escaped the way it had come. The big bird, oblivious of an audience but clumsy in a small space, crashed into the glass of a closed window. There was a horrid thump, then it fell to the floor and lay still.

I looked at it curiously. The yellow eyes were fixed and staring, the body limp. Surely it was dead. I picked it up and the others all gathered round. Gently stroking the smooth, dark feathers on the back, I turned the bird over to admire the speckled feathering on its breast. Fast, fluttering movement! The heart was beating! All at once, powerful talons were gripping my thumb and beady eyes regarding me with a distinctly malevolent expression. I ran to the door, yelling with the pain, and raised my arm into the air. The casualty took off as if nothing had ever been wrong and flew straight for the nearby wood.

I felt a disapproving presence behind me. Mr Robertson had come to check on his missing pupils.

'What was that bird, sir?' I asked eagerly, full of the exciting experience and curious.

He shrugged indifferently. 'I have no idea. Hurry up. You're late.'

That was my first sparrowhawk.

We spent most of our time in school planning what we might do once the day was over, or during the holidays. Catching rabbits was one ploy, though really it was best done late in the evening or early in the morning. They made good eating and our mothers were happy to stew them. The weapon used was a catapult made from the fork of a hazel branch and a strip of rubber cut from the old inner tube of a tyre. I once managed to stun one! Fishing was another great interest and us boys were lucky to have a mentor, Harry-the-Englishman, who had two thumbs on each of his hands and was renowned for his capacity to down large quantities of beer. He was happy to demonstrate with a greenheart rod of his own which he treated with loving care and would lend to nobody. Through the Stewart tackle on the rod he would run out a long line with a worm for bait. This he would expertly cast to a likely spot in the river, then move it up and down cunningly with the current while at the same time steadily making his way downstream. Our rods, however, were not greenheart. They were bamboo stems stolen from the gardens of the Estate. We became quite good fishermen and often took home a catch for tea.

In school one day, during the lunch break, the word went round that 'Sandy' would be killing a stirk that evening. Sandy Smith was the butcher. He killed, cleaned, and butchered all the meat for Ramsey's shop in Front Street, and served in it, too.

'Are you coming?' asked my friend Alan, for the butcher had no objection to an audience.

'Well, alright,' I replied, somewhat reluctantly. I did not enjoy watching anything put to death but did not want to seem a sissy. Anyway, it was interesting in a way.

The small slaughterhouse was right behind the shop but the pathetic mooings or baa-ings of captive beasts never seemed to worry the shoppers. The heads of the bullocks were drawn down to the floor by means of a rope through a ring. Then they were killed with a blow to the head from a wooden shafted weapon, a hollow spike on the end of it. The sheep had their throats cut, the blood being saved for black puddings. Later on that day, I chickened out, finding an urgent job my mother needed doing.

Katy Cameron presided over Ramsey's, which also sold a few groceries. She seemed blissfully unaware of the flies and blue bottles which settled in their thousands upon her wares, the steaks, gigots, mince, offal, bacon, sausages, and so on, which were displayed in dishes on the counter. Fly papers were always full of dead or squirming victims and if there were only one or two shoppers, the sound of buzzing was louder than the gossip. Yet, nobody died from food poisoning!

One day, an after school prank altered the course of my young life and also put my father's plans for the future in jeopardy. We went turnip stealing in one of Archie-the-farmer's fields and laden with our filthy booty repaired to the Big Burn to wash it. It had been raining all night and foaming white water roared through the meadow on its way to the Add. I should have known better. Leaning forward too far, the bank gave way. I lost my balance, fell in, and the shock of icy water, or a blow from a rock, knocked me out. Next thing I knew, an alder branch was whipping across my face and sheer instinct had me grabbing it. Scared friends hauled me up the bank. For a few minutes I lay gasping, unable to speak, then the shivering started. Run for home? No way – I should not have been anywhere near the burn. I stripped off my dripping clothes, wrung them out as best I could then replaced them with difficulty. For a miserable hour, I ran about trying to get dry and warm but, at last, still soaking and long past caring whether there was a row in the offing, I slunk home. The consequence was a double pneumonia which nearly killed me.

As a result of this escapade, a lot of schooling was lost. But my father was not dismayed. I must work hard. There was still time. He kept my nose to the grindstone and, in due course, I won a bursary to the Grammar School at Dunoon, a school with an excellent reputation for getting its pupils into university. It was a huge and exciting change. First, the school itself seemed enormous – many teachers, many classrooms, many pupils some of whose names I would never know, and, of course, a much broader curriculum. It was far beyond a daily trip in a bus from Kilmartin, so I boarded with a local family, returning home only for the long holidays when term ended. This experience is a common one to most Highland children and is one which they mostly take in their stride. However, it is also, perhaps, the sowing of a seed, the idea that in order to succeed in life you had to leave your native 'place'. Once settled down, I thoroughly enjoyed life. I was standing on my own feet, fighting my own battles, and even working quite hard as well. At the beginning of Fourth Year I was actually judged to be university material. My father was pleased and on the rather flimsy evidence that I was good at dock leaf repairs to punctures in the tyres of our old bikes, planned that I would obtain a degree in civil engineering.

Great. But then, towards the end of that year, I went down with a bad attack of rheumatic fever and that was the catalyst that ruined it all. I was sent home to be nursed and the doctor became concerned for my heart. My distracted mother took me to a specialist in Glasgow and there followed a long year in the Western Infirmary where the only cure available was complete rest.

Fuming, furious, and terribly frightened at first, I just lay on my back, bored out of my mind and scared by the inevitable stress of being in a mens' ward where some were in pain and dying. Charming young nurses fussed over me. A well-meaning dragon of a Sister bullied me and checked the size of my greatly enlarged heart once a week – this done by listening to its irregular beat and tracing it in indelible ink on my chest. Its gradual shrinking was duly noted and recorded. Endless months later, I was able to sit up for half an hour and eventually, three stones lighter, hardly able to walk, I was sent home to recuperate.

This was the time of the great depression! I was incredibly weak and if I recovered my strength at all, it would clearly not be for a long time. The doctor shook his head and advised lots of rest, no strenuous exercise and perhaps a job in an office one day. My mother was over-protective and nearly drove me mad. My father, however, still ambitious and quite undaunted, soldiered on. Sure, I had missed lots of schooling, but I could do Sixth Year studies at Kiel, a 'public' school in Dumbarton, to which I would win a bursury and thus obtain the necessary qualifications to take me on to university. What a hope! In due course and inevitably, I failed the Entrance Exam and from then on he lost interest and mostly ignored me. Gradually, as the long months passed, I began to realise the full implications of my illness: a damaged heart, an incomplete education, no qualifications, no career. What was I to do? Be forever dependent on my parents? Live on the dole? I began to feel written off and of no use to anyone, especially myself.

A young adolescent, impatient, and horribly sensitive about my condition, I must have been a pain in the neck to one and all. But everybody was incredibly kind and I did improve. A total inability to sit still indoors was probably a help and short walks about the village soon became rather longer ones. Dugie-the-Post, red-haired and always cheerful, encouraged me by sometimes allowing me to deliver his mail in the village – he even paid me the generous sum of 10/- to do it. Then, in the spring, a miracle occurred. It was called Eilean Righ and was a small rocky island in Loch Craignish. Two miles long and barely half a mile wide, it belonged to a recluse, a professor of Chinese history and philosophy. The old house had no telephone. Telegrams and mail had to be delivered from Kilmartin on foot. The walk was long but involved nothing too strenuous: a gentle climb to the Lady's Seat, a nearby hill; a saunter down a long, narrow glen, with steep cliffs on either side and finally, a stroll through oak and birch woods to a rocky coastal plain and the jetty. From there, they were rowed over to the house. I decided this was for me and, Son of the Post Office, it was agreed I could try.

There was a bonus, too. During my Dunoon school years, my interest in natural history had suffered due to lack of time. Now it came into its own. Few people ever walked to Eilean Righ, so there was almost no human disturbance. Rabbits abounded and foxes, stoats, a pair of buzzards and even the elusive wild cat, though I only ever saw its droppings, made use of this prey. I learned to come silently, to keep my eyes open and sometimes just to sit quietly, hidden behind a boulder or a bush, waiting for something to happen. There was a lot to discover and all the time in the world to do it in.

One morning, I was perched on a rock some way above a rabbit warren on the slope below the Lady's Seat. It was a great day and going to be warm. A small movement at one of the holes caught my eye and turned into a twitching little nose testing the air for scent. In a few moments, a long, lithe, low-to-the-ground creature crept into view. A stoat. The tawny brown fur on its body glinted silkily in the sunshine and blended into the warm, soft cream of its throat and belly. The animal was looking back over its shoulder. Five smaller versions came bounding to join it. Then all, mother and children alike, began running about busily, noses to the ground, evidently looking for scent. The youngsters were never far from the mother, though, and kept checking that she was present. All the time, I could hear them chattering together with low-pitched, trilling, companionable sounds that were not aggressive. The kits, like their mother, had black tips to their tails but their fur seemed paler and softer than hers. All looked sleek and well-fed.

All at once, the mother was off, nose down, along the hillside. The kits bounded after her, nipping, scrapping, dancing along behind, falling over each other in their anxiety to keep up. The mother streaked into another hole and with no hesitation at all, the family followed. And, that's that, I thought – they'll not come out again. But then came the panicked squealing of a rabbit. Moments later, the adult appeared with a young one gripped in her jaws. The kits charged towards her, chittering with excitement, a fierce little mob all trying for a bite. But she did not hand over her prey. Reversing down the hillside, she dragged the furry bundle, all torn and bloodied, through the vegetation, over a stony patch and even round a rock that was in her way. All the time she was hindered by her ravenous family. At last, she came to big boulders, a jumble of enormous rocks at the bottom of the slope. She pulled the rabbit after her into a cavity beneath one of them and the rabble followed her in. I did not see them again. Perhaps that was their den.

I learned a bit about stalking on my walks to Eilean Righ and started, in earnest, with the Common seals which were usually found basking on the island skerries and the mainland rocky shore. It needed superhuman patience and some skill to get close to them and, at first, I was no good at all. Beautiful creatures, with soulful eyes and a gentle, inquisitive expression on each face, they possessed scenting and hearing powers of great sensitivity. By the time I was poking a cautious head round a boulder to take a look, their great lumbering bodies were already humping over the rocks to the sea. From there, enquiring heads would bob up to take another look. It was useful experience for later when I would stalk these creatures, along with many others, for photographs.

Down at the ferry one day I noticed a heron, with full crop and labouring wings, flying north along the coast. As I waited for the boatman, three more flew overhead and he confirmed that there was a heronry in an old aspen wood near the shore. It was only a couple of miles. I ought to look at it. There was no urgent need to return home and it was a super day, crisp and dry with a northerly breeze. I set off on an easy walk over springy, close-cropped turf where little groups of sheep were grazing sand-sprinkled grass – they glanced at me with incurious eyes and moved, seemingly reluctantly, only yards out of my way. Occasionally, I dropped down to the shore to firm sand and a carpet of shells. There, where a burn filtered through to the shore, I found a dark dropping and prints of webbed feet. Later, my reference book confirmed an otter must have been there. Eilean Righ lay behind me, its old mansion house brooding and lonely. Ahead, the rocks were smooth from long poundings of the Atlantic and, in the far distance, hills shimmered in golden sunshine.

An overwhelming stink of rotting fish suggested the heronry must be close. I turned into the awful smell, looking for aspens but finding instead a small birch wood. I followed my nose and, walking as quietly as possible through the smooth stems, came at last to a small clearing of flaming gorse bushes. There, on the far side, stood the wood I was searching for, its trees tall and slender, their leaves fluttering silver in the breeze. In their branches nestled great baskets of sticks. A heronry, indeed. The only sound was the whispering of leaves in the canopy, so if there were any chicks they must all be asleep.

Not sure how cautious I needed to be, I walked slowly and carefully towards the little wood and stopped on its edge. A good thing. Pausing to check the best way to go I spotted a fox only twenty yards ahead. Bright rufous in the sunshine, it was sitting on a carpet of broken blue eggshells and was eating something. Feathers and down were scattered all around and its prey looked remarkably like a young heron. With apparent enjoyment, a long tongue lick-licking, sharp teeth tearing, a jaw working hard, Reynard continued his meal. For nearly five minutes I watched him. Eventually, nothing useful remaining, he, for it was a dog fox, rose to his feet, shook himself vigorously – a shuddering of red fur from neck to tip of long thick brush – then began to potter over the woodland floor. From tree to tree he padded, nose down sniff, sniff, sniffing, until suddenly he paused beside the stem of an aspen. In a second, he was on his hind legs, stretching eagerly upwards. Nose twitching, ears pricked to catch any sound, forepaws scrabbling to reach ever higher, something held his interest. A huge nest seemed to be the focus of his attention and it was certainly occupied for its sticks were all covered in 'whitewash'. So that was it! Perhaps his prey had been a chick fallen out and now he was looking for another. Would he climb the tree? I knew he could.

At that moment, from above the wood, came the steady beat of powerful wings and I caught glimpses of several large birds. They were herons. All at once, there was cacophony, a sound like the looms in a spinning mill all rattling away – chicks, all calling to be fed. My fox thought it time to go. I watched him slink casually, not too bothered, not really hurrying, through the trees towards the crags behind the aspen wood. He disappeared into the dark shadows of boulders there, and was seen no more. At each nest, one by one, a huge grey bird with fast beating wings braked furiously, then dropped to its untidy platform of sticks. As it clumsily touched down there was sudden silence and I imagined pale yellow chick bills reaching up to fearsome orange adult ones to gobble and gollop regurgitated fish. I waited until the adults had departed again for the shore, then crept away.

So the weeks and months passed. I felt stronger and enjoyed each one of my walks to Eilean Righ, whatever the weather. But feeling fitter also meant being restless once more. Was I going to run errands for Dugie and take the mail to Eilean Righ once or twice a week for the rest of my working days? Was I to be stuck in Kilmartin? Just to earn some money and do my injured ego a spot of good, I took a skivvying job in the house on Eilean Righ. That was a mistake. I was at the beck and call of all the other servants and in my rare free moments was too tired to do anything but sleep – no wildlife watching there. Then, the following summer, it was off to the *Duchess of Hamilton*, a turbine driven ship which had recently replaced the old paddle steamer which sailed on trips for trippers on the west coast. Here the galley heat was almost unbearable, with a deck so hot bare feet could not stand on it, endless food and pots of tea to be pushed through a hatch to the saloon, and unlimited washing-up. Not for me. Whenever we sailed up Loch Fyne, on our way to Inveraray, I looked at sunlit hills and woods dappled in sunshine and shade, or even the same shrouded in mist or rain, and wondered what the hell I was doing on this old tub. I returned home. My father began to mutter about 'that job in the bank the doctor could probably get me'. No, thank you. By now I knew more than ever that clerical work in an office had no attractions. I must find something else.

At that time, my father was employed as a forest worker by the Forestry Commission. Suddenly, out of the blue, the thoughts came chasing through my mind: that is work I can do; I'm fit enough now and that way I'll not be stuck in an office. Both parents thought I was mad but, the decision made, it all happened quite quickly. The forester-in-charge agreed to take me on and almost before I could believe it, early one morning in November, I was joining a group of oilskin-clad men on the old jetty at Ford. Ford was a small village at the west end of Loch Awe and the forest reached roughly eastwards for mile upon mile along its northern shore. The squad, I had been told, was working near the far end and, no road available, they travelled there by launch.

I reported to Drew MacFarlane, the ganger, or foreman, who handed me over to a chap called Johnny. He would tell me what to do when we arrived at our destination. Then I took a look at the *Mary Ann*, all forty feet of her, and the loch stretching away into the distance. Bitterly cold, the snow already on the twin peaks of Cruachan, a brisk breeze from the north was churning the water to choppy waves. I loathed water, was afraid of it and could not swim. I shivered. I had not bargained for this. The men, however, seemed a friendly crowd and, of course, they knew my father, though he was in a different squad. I recognised Harry-the-Englishman and knowing well how I would be feeling about a trip in a boat, he gave me a fiendish grin with a thumbs up sign. As we pulled away and began to rock with the swell, good-natured teasing came my way: 'How you doing, Don? Feeling sick? Downwind of us, please. This is nothing. Don't fall in. Have some chocolate.' Oh my god!

The skipper was called Big Dugie. 'Big' for obvious reasons, for he was tall and burly with a magnificent red beard and a fresh complexion to match. A favoured few were allowed to join him in the cockpit, but the rest of us sat in the open hold turning our backs to wind and waves. Stomachs were strong! Some of the men were smoking pipes, a few had cigarettes, and one was eating a late breakfast 'piece'. A man sitting quietly apart wore a deerstalker and held a rifle across his lap. I assumed he was the trapper, or stalker. We kept lurching sideways, the skipper unable to hold her on course, and it seemed the whole loch came pouring over the side. How long would this journey take? How far did we have to go? But as we chugged along, oily smoke belching from the ancient engine, the bows cleaving a path through the waves, I gradually got used to the awkward motion and began to feel better. We were sailing close to the north shore. First impression there, through the flying spray, was of an autumn-tinted fringe to the lochside of oak and birch and behind, a forest in deepest green and with the gold of larch marching seemingly for ever over the hillside.

Half an hour later, the launch headed in towards the shore and soon our unwieldy, flat-bottomed craft was crunching to a halt on a small pebble beach. It was a pleasant little bay, the water shallow on a sandy bottom. With stiff, half-frozen legs, we climbed over the side and waded ashore. A curious, ramshackle contraption, ugly and functional, stood on a steep bank above the loch. There was a pile of logs beside it. Johnny told me it was a chute down which the logs were shunted to the loch. Then they were taken by launch to Ford for onward transit by road.

'What are we doing, today?' I asked him, eager to get to work.

'Thinning.' His reply was short but as we climbed the bank and marched up the steep track through the oak wood to the forest, I saw myself bringing conifer giants crashing to the ground, their days brought to an end with the chopping of my axe. Not quite.

'Right,' said Johnny, when, at last, we halted in a clearing surrounded by trees already cut. 'We're humphing. Watch me.'

While the others were preparing for work, he led me to a fallen tree already freed of its branches and cut into six-foot lengths. He shrugged off the bag he carried on his back, removed his jacket revealing a filthy old sweater, ostentatiously flexed his biceps, and got down to work. A small, dark, wiry man, he lifted a substantial log to his shoulder without apparent effort, shifted it into position with a grunt, then stood waiting for me to do the same. Easy enough, I thought, rolling up my sleeves. I picked the next and tried for a grip. Impossible! It was slippery from the rain of the night before. I shifted my hands a few inches, thought I had it safe and with a deep breath, attempted a lift. The damn thing wouldn't budge and seemed nailed to the forest floor. I tried again but once more had a job getting a hold. I couldn't shift the brute. I straightened up, took a quick look in Johnny's direction expecting to find him in fits. But he seemed intent on the way he would go. Good. This time. With an awkward struggle I, at last, brought the log to the level of my chest. Bloody heavy. Took a deep breath. Couldn't hoist it higher.

I dropped the damned thing and skipped smartly out of its way.

My mate, now, was nearly killing himself.

'Take it easy,' he spluttered. 'It won't go away.'

Next time round, I made it. The monster log was safely transferred from ground to shoulder and then balanced precariously. Johnny was already away. As I set off to follow him, sweat pouring down my face, knees shaking, I felt a complete idiot. He strode along with an easy rhythm, his log held steady, but I tottered after. With every stumbling step my load became more cumbersome, uncomfortable and uncontrollable. Twice it slid to the ground and was hauled, with difficulty, on board again. As I staggered down the path, I thought: this job is not for me, I won't even last the day. But when, in due course, we reached the chute, shunted our logs into its gaping maw and heard the splash as they reached the water, I felt a certain satisfaction. It would be easier next time.

In our first break, Drew explained to me that the trees were Norway spruces and those cut would be going to the mines for pit props. I watched one falling – creaking protest, fibres tearing apart, branches swishing through the neighbouring trees. It thumped to the ground and two of the gang started snedding the branches away, quick, accurate, energy-saving movements with an axe. The stem was then measured, cut into lengths and carried to the lochside. Humphing! With a break for lunch and another short one for a mug of tea, that was my day. Heave a log up, balance it, shift it, balance again, stagger to the loch, despatch it down the chute, and return for another. By lunchtime I was no longer admiring the beauties of nature nor the expertise of my fellow workers. I just existed in a sea of pain. When at last I limped to the launch, my neck and shoulders were aching and raw, my legs were weak and my heart was thumping uncomfortably. However, I was still alive, the loch was now calm and the journey smooth and Johnny said it would be better the next day.

It did not take too long to settle into the new routine. The early morning ride past the Bull Rock to Ford was exciting, for bicycles, even my old Post Office cast-off, were a silent means of travel which did not disturb any wildlife that might be about. It was surprising how often a fox, rufous and bushy-tailed, would be trotting down the road on its way home for the day and twice I saw a stoat, its coat turning to winter ermine, flashing in front of me and diving into the hedge. The peregrines were more difficult to see, the breeding season now over, but there were buzzard, kestrel and sparrowhawk to watch out for. Trees began to interest me, too. Why were conifers the only trees being planted in the new forests? Why not deciduous trees, as well? How were these forests planned? And so on. Until now I had lived among the birch and oak woods around Kilmartin. Now a vast new kind of forest was mine to explore.

For a year or two, however, I had neither time nor energy for much in the way of exploring, and wildlife forays were mostly looking for tracks and signs during our brief lunch breaks or on weekends at home. We were busy in the forest and there were new skills to learn. After humphing came digging the drains for a new plantation in Knapdale where a vast new forest was taking shape, grinding work for there was no mechanisation, though it was soon to come. The area had already been fenced to exclude those avid consumers of young sapling shoots, sheep and deer, and we were to make it ready for the planting of the trees. We worked in groups, four in each, spaced across the hillside. Two men, each wielding a 'rutter' (a right-footed spade, or left-footed as required by the person wielding it), worked across the ground digging v-shaped wedges. Another man, with a cross-cutting spade, followed behind slicing across the 'v' so that it was cut into sections. The fourth member of the team then dug out the turves with a three-pronged fork called a 'hack' and threw them into two rows on either side of the drain – thus forming four rows between each. The ground, with a high moisture content, was heavy, peaty and sticky. It was winter, the weather often awful and the work backbreaking. The job took weeks to complete.

Then came the planting. That was in the spring. I thought it would be pleasant work after the draining and did not believe Johnny when he told me that we 'only' had to plant a thousand in the day. A thousand! It must be an easy job.

By now, more used to the vast areas of ground that would become a forest, I was not too dismayed by a seemingly endless hillside, all drained and in neat order, reaching to the ridge and east and west as far as the eye could see. Everyone else seemed to be taking it all for granted – the job would be done quite quickly. We divided into groups of three. Bundles of small trees had been previously dropped into drains at strategic intervals and we spread out to match them. In my group, George was the 'splitter', his job to go ahead and cut a slice in each of the turves from the centre to the outside edge. He set off straight away. Johnny strapped a bag around his waist into which he placed a bundle of trees from the nearest 'sheugh', and told me to do the same. Off we started after George. At the first turf, Johnny quickly lifted the loosened sod, popped a young tree in by its roots, replaced the sod then firmly stamped on it. It took seconds, only, and in no time at all, he had planted the next. And, so on. It was a lovely day, sunshine, but not too hot and with a gentle breeze to keep us cool. Wheatears, meadow pipits, wrens and robins were all sweetly singing that spring was here. It felt good. This would be pleasant work. But in only ten minutes, with an easy rhythm that took him along surprisingly fast, Johnny was way ahead of me and George had already started on the next row above. I laboured along behind, over anxious, taking too much care, and swearing at the prickly little saplings as I pulled each one from the bag. By the end of the day I had managed 800 trees. All the rest had clocked up 1000 and some many more.

Over the years that followed I did all the jobs that forestry workers were required to do: draining, fencing, planting, felling, maintaining forest roads, making birch beaters for putting out fires, and doing various other odd jobs as well. Never got to fell a tree, though. There began to be a little more time for wildlife. Chris Petrie, the stalker, sometimes took me along on a Saturday morning, Drew the ganger turning a blind eye to my absence. The stalker took a lift on the *Mary Ann* to the top of the loch and then we worked back, against the wind, down through the forest. My experience with the seals at Eilean Righ proved useful. I soon learned to track deer until I could approach quite close and at that time, I gained satisfaction from a clean shot from Chris which killed instantly. I never wanted to do it myself, though.

On one of these occasions, I picked up a curious dropping, almost black, full of little bones, smaller than a fox's.

'Ever seen one of these?' I asked.

'No,' the stalker replied, thoughtfully, but seemed quite excited. 'It might be pine marten. I didn't think there were any, here.'

I felt quite chuffed, but we never actually saw any of these animals nor any more of their scats. Perhaps one had just been passing through.

By now I was really fit, except for the odd blip from my heart. But, gradually, as each day passed, the same slog with minor differences, the rest of the chaps seemingly content, I began to be restless once more. I suppose I needed a challenge and some prospect of improving my lot. In the evenings I read a lot and forests, old and new, began to fascinate me. I suddenly realised that the bare hillsides I was used to were man-made, the result of centuries of exploitation of the timber which once was there. I learned that the Old Wood of Caledon, that marvellous forest which used to cover much of the Highlands, had all but disappeared. What had it been like? What wildlife existed within it? Where were these remains? I thought of the new forests. Could they, in any way, replace that old pine and birch. Certainly not, if they only planted spruce! But new forests could provide a habitat for wildlife and improve a landscape that could easily, in the centuries ahead, become a desert if left un-nurtured. All that in addition to the timber they would provide. So I dreamed.

One day the thought occurred. It would be great to be involved with the planning of a forest, to know the thinking behind it and to be putting those plans into effect. Could I somehow, in spite of my lack of schooling, become a forester, someone with responsibility for the job that would be done? What to do? One evening I phoned Bill Hendry, the forester-in-charge, and was invited over to see him.

Want to be a forester?' he asked with a smile and did not seem particularly surprised. 'I can certainly recommend you for Benmore. You ought to pass the entrance exam all right.'

three: coillessan

BENMORE was great. From the first reverential walk up its magnificent avenue of tall Sequoias to the moment when I reluctantly had to leave, I enjoyed every minute. The college, taken over from its previous owner, was an old stone-built mansion in a dramatic setting of rugged mountain tops and crags. Beautifully mown lawns surrounded the building and merged with mixed woodland of oak, ash, birch and hazel. Dark pine, larch clad in freshest green, and spruce, a different green again, climbed the hillsides above to heather moorland and outcrops of rock. A hurrying, scurrying river, rock-dotted and rock-pooled, flowed through the grounds from Loch Eck whose waters sometimes danced in the sunshine but, more often, were sullen and forbidding in overcast conditions or ripping and roaring in a stiff northerly breeze. Beyond were the mountains of the north.

The course was a mix of the practical and the theoretical. Each morning we set forth for the College's own little forest to learn the skills of draining, planting, felling and so on. It was mostly old stuff to me, but I now looked at it as one who would plan and give orders. The afternoons were for lectures and these, basically, were concerned with silviculture, the growing, and the conditions in which they could be grown, of the various species of tree that were planted in the forests. In each area the soil would have to be tested and exposure to the prevailing winds noted, so that the species which were most suitable would be planned for each site. Thus the end product would be a nice mix of pine, larch, fir and spruce. The whole subject fascinated me and I could envisage, in years to come, forests of varied species, full of dappled light and space, where people liked to wander and refresh their jaded spirits. So I decided! The economics of forestry were forgotten and, for the present, I saw no further than the diploma at the end of my course which would enable me to have a part in creating this paradise. A career assured, a ladder of promotion to climb.

After only a year, the Second World War broke out. Benmore was closed for the duration and we all rushed off to join the Army. At least, my colleagues did. Much to my dismay, I was rejected because of the damage the rheumatic fever had done to my heart. I was shattered, depressed, and horribly sensitive to the fact that my friends were all going off to the war and I was stuck at home. Worst of all, I felt so fit. How could this be? So it was back to the forest once more, the fencing, draining, and planting of new forest and the thinning of the mature. Strenuous body-building work and the irony that, supposedly sick, I survived it all.

Benmore did not re-open. Four years later, after the war was over, I found myself instead in the Forest of Dean to complete my training. The rest of the Scottish contingent was also there and we were joined by those who had already been in training there before but had gone off to the war. A number also came from the universities, preferring to do a straight forestry course rather than to obtain a degree. Now the pressure was really on, a shortage of timber in the country and the growing of trees an urgent priority. We did no practical work at all, attended a great many lectures, sat endless exams and completed what should have been a year's course in six months. Much to my surprise, I graduated top of the class. It has to be admitted I was pretty pleased with myself and thought I knew it all.

And then, to my first posting at Ardgartan Forest. Here I was, cycling home from my first interview with the Head Forester angry, frustrated, even considering a request for transfer to a forest where my newly acquired skills would be appreciated. But the day was still good.

Sunshine bathed hill and lochside in a golden glow and when my new home came in sight it appeared old, sadly neglected and asking to be lived in once more. My spirits began to lift. Maybe I had misunderstood Jimmie, who might well have had more weighty matters on his mind than a new recruit. Did I not want to live and work in a solitary place? This one could hardly be bettered. All right! Tomorrow I would go down to Mark to see Old John and explore a part of the forest on the way. Today I would put my house into some sort of order.

How to get to Mark? First thing next morning, I took a look down the loch through binoculars. Jimmie had not given me a map, a deliberate response perhaps to a cocky young man who should have known to ask for one. All I did know was that the old settlement lay some six or seven miles away and that there was no road beyond the dirt track which ended at my house. It looked like a bit of a challenge. Forest of varying ages, Norway spruce, European larch and Scots pine, stretched for ever into the distance. The further away, the trees were younger and beyond them still, I could just see the parallel patterning of land already drained and prepared for planting. In the foreground thin pencil lines, etched on deep green, suggested the 'rides' which divided each plantation from the next and also broke them into manageable sections. They were not obvious in the dense growth of young trees and I suspected would be difficult to find once I was in among them. No obvious route that way.

I turned to examine the shoreline. It was bordered by the rusts, golds and yellows of deciduous woodland which spread in colourful ribbons up gullies on the hillside, relieving the monotony of conifer green. A storm-torn zone separated forest from loch, boulder broken and pebble strewn, a mosaic of seaweed-covered rock and corrugated sand. A number of pine and heather-clad outcrops poked rugged noses into the loch and high water mark on the first warned of problems for the unwary. That, however, seemed the way to go. It ought to be reasonably easy and Mark should soon be in sight. What about the tide? I watched wavelets lapping and licking over sand and seaweed and decided each succeeding one did not climb so far up the shore. Safe enough.

Full of determination to prove myself to Jimmie, I set off into a crisp clear morning across the burns of the Coillessan Wood. The loch was as yet darkly still, reflecting the sombre velvet of the forests. On the hillsides, a golden sun already warmed bare birch branches to rosy brown and the dull sheen of spruce to a fresher, deeper colour. Nearby, on the shore, the rocks sparkled with frost, pitted pale grey glinting with a myriad diamonds. Springy turf was pleasant walking but once beyond the Wood of the Waterfalls the ground became increasingly soft. Dead leaves of oak and ash concealed a woodland floor covered in mosses, especially sphagnum, a bright green carpet which hid squelching, heaving bog. I was soon in trouble.

The stems of three rowan trees, still holding the berries of autumn, had been almost stripped of bark – I knew the culprits were the deer who browse on the bark of these trees and seem to find some special nutrient therein. They must be coming down from the forest, for a track disappeared off towards it. Follow that too, I thought, and there'll be drier ground. There was, but I soon met thickly planted spruce through which there could be no easy way. Back to the bog. It was while I was cautiously treading yet another track that it suddenly occurred to me that perhaps it had not been made by deer at all, but by some other animal. Narrow and kind of unobtrusive, it wandered secretly through the wood, past the stems of the old oaks, round and between moss-covered boulders, and eventually seemed to be making for the shore. Then, a clue. The track disappeared beneath the undisturbed branches of myrtle bog and emerged again on the other side of the patch. No deer could pass through that. It had to be a low-to-the-ground sort of mammal, I reckoned. I found a dark dropping and remembered the one old Harry-the-Englishman had shown me so long ago. Otters! I still had never seen one. Perhaps I would surprise a family on the shore.

Otters or deer, I still kept sinking into the mire. Hard work and time-consuming. Better make for the shore or the job would take all day. A few minutes later, I was looking at dark-swelling waters rippling in from an immensity of blue distance to caress sand, seaweed and rock, polishing them to fresh radiance in the sunshine. To my right a confusion of jagged rocks built to outcrops topped with heather. And then, backdrop to all the deciduous woodland in autumn array, close-knit conifers climbing the hillside to the ridge above. Yesterday's snow had not

vanished from the mountains and I wondered if the coming winter
would be long. For a while it was hard work over soft sand but then,
thankfully, back to dry land and level stretches of bleached grass over
which I made good time. I began to notice droppings all scattered
around. Sheep! Did Old John's marauding flock wander this far?
Perhaps both sheep and deer would be on their way to the seaweed lower
down the shore. I knew they liked to feed on it.

A buzzard surprised me, lifting silently from the branch of a grand
old oak. As if bound for the other side, it flapped slow and steady over
the loch and its plaintive mieuing emphasised the loneliness of the place.
Unseen curlews called wistful messages to each other in a faraway
distance. A flight of oyster catchers flashed black and white as they
speeded up the loch then whistled back again a few minutes later.
Whether they found feeding grounds unsatisfactory or were just on a
morning patrol, I could only guess. In one little bay, riding a gentle swell
and diving for a meal, I found an eider duck, drab in brown feathering,
with three handsome drakes in attendance.

More alert now, I picked out a pair of eagles to the west, high above
needles of rock which pointed jagged teeth into the sky. Leaning on the
air with barely beating wings they floated effortlessly on thermals over
the ridge. All at once, one came plummeting down, stooping at
incredible speed towards the ground. The other, dipping a casual wing,
followed after. Both came hurtling earthwards as if bent on suicide and
I wondered if they were after a rabbit or hare. But no. In only seconds,
they were climbing steadily to the skies again and neither held prey in its
talons. For a long moment they circled, majestically gliding, wings
dipping to change direction, rudder tails compensating. Then, suddenly,
in rays of sunshine light, heads down, golden bodies almost still, they
were hovering. Prey below? All at once, one broke away. Swift as a
peregrine falcon it came streaking down and the other came racing after.
Number Two curved gracefully beneath the falling body of its mate,
then accelerated away, spiralling effortlessly back into the blue. Number
One, still on a downward course, looped a graceful loop then soared
after, to overtake. For a second two eagles, in perfect symmetry flew side
by side. Then, one after the other, in another headlong flight, they

hurtled towards the heather, alighting in a corkscrewing, wing-fluttering, last second braking-escape from death.

It seemed like sheer high spirits, a joyful moment in an eagle pair's day. Unfortunately, they were gradually edging further and further north in the cloudless sky and soon were lost behind the ridge. Perhaps their eyrie was somewhere up there.

For a while I had to keep entirely to the shore. I crunched over pebbles, scrambled over rocks, and found stepping-stone passage over the estuary of each little burn. Every so often a heron, standing in the weed and staring with total concentration into the water, rose with noisy protest at my coming and with ponderous, hard-beating wings flapped off to find a new stance elsewhere.

Most of them settled again within binocular distance and, with a deceptively casual air, began patiently waiting for the next meal – a lightening strike with a dagger bill was success. A mile or so further on, I discovered why there were so many of these birds. Above the shore, hidden by taller oak and ash and bordered on the north side by mature spruce, grew a small copse of aspens. Their remaining leaves, fluttering bravely silver in the chilly breeze, could not conceal the untidy platforms of sticks resting, apparently precariously, in their branches. Each giant nest was bespattered with droppings. A heronry! Just like the one at Eilean Righ, but smaller.

By now, of course, all the young had flown but they were still around, perching in the surrounding trees and squawking disapproval of my presence. One by one, they took off as I wandered in among them.

The heronry marked the end of easy walking. It was the tide. I had got it badly wrong. It was racing in. A hasty glance at marks on the nearby rocks indicated rapid retreat. I scrambled to safety, but over the next rise found more spruce, thick and seemingly impenetrable, growing right down to the shore. This is ridiculous, I thought. People must go to Mark. Did they walk there only when the tide was out? What about Old John? Did he never go to the village for supplies? And so on.

There was, now, only one possible route. A ride which would take me through the trees and up to the open hillside. From there it might be possible to locate John's cottage at Mark and plan a way down to it. But how to find one? Looking at the gradients involved, the thousands of young trees growing close together with no obvious path through them, and then to the heather and rock above the forest, I knew that the going would be tough. Thank you, Jimmie. With bad judgement and a lot of impatience, I plunged into trees about ten years old, planted the regulation five feet apart; drains, running a parallel course across the hillside, bisected each row.

In no time at all, I was in trouble. Branches, thick on the stems almost to the ground, intertwined with the next tree, with seemingly fiendish intent, to hinder all progress. Branches whipped across my face to threaten eyesight. I tried a travelling backwards technique so that my rear end and back would take the brunt. That ploy ended in an undignified collapse into a deer wallow no doubt made by an ardent stag – the rut was only recently over. I tried walking one of the drains, hoping it would bring me to a ride. That was disaster, too, for I was soon forced to crawl on hands and knees – a needle-bottomed, branch-canopied, rock-strewn way, but the only practical one. I battled on, pushing, shoving, swearing, from time to time rising out of a ditch to barge over to the next one alongside, and so make some progress up the hill. It was an endless nightmare. Scratched and bleeding, my knees bruised and my back itching with myriad needles beneath my sweater, I knew I was an idiot. But there was no going back. I doubted if I could, anyway.

At last, lifting my head cautiously to see if there was light at the end of an excruciating spruce tunnel, I glimpsed an expanse of paler green ahead. It could only be a ride or a small clearing in the forest. In a few more yards I was wearily unwinding my body to an upright position and dragging each protesting foot from the mire. Miraculously, a glorious vista of green, broad and tree-lined, marked a way right up the hill and the ridge was clear above it.

I began stripping off sweater and shirt and was busy picking the worst of the needles from my back when a dog barked sharply from somewhere not far away. Dog almost certainly meant man. Who could possibly be walking in this remote and uncomfortable place? I was hurriedly replacing my clothing when a small terrier came racing out of the forest. Hot on my scent, he drew up at my feet with a most unfriendly growl.

'Hello, boy,' I said cautiously, keeping my hands firmly in my pockets. 'Where've you come from?'

Then someone whistled and a man dressed in old tweeds and a dilapidated deerstalker came striding down the ride towards me. He was short and lean with a weather-beaten face and a dapper little moustache. He carried a rifle and walked with the energy-saving, easy gait of a man who covered long distances on the hill. I guessed he was the forest stalker.

'Come away,' he ordered the animal quietly. Then, with hand outstretched: 'You must be Don MacCaskill. Alec Seaton. We thought you were poachers!'

'Poachers?' I exclaimed, though I certainly must have looked like a particularly scruffy one.

'Aye. They come in from the loch. Are you all right?'

He was grinning, in a friendly sort of way, and I reckoned he knew what I had been doing.

'I'm fine. Jimmie Reid asked me to go and see Old John down at Mark,' I went on to explain, embarrassed and feeling I did not exactly look like someone in authority. 'The shore seemed the best way to go but the tide is coming in fast.'

'You're joking,' Alec exclaimed. 'We go by launch to Mark. The boat is used when there are no roads.'

'There's certainly no easy road to Mark,' I commented wryly.

'The *Mary Ann*'s out of action at the moment. He's an old bugger sending you this way.'

'I did forget to ask for a map,' I admitted.

Alec looked mildly amused and it dawned on me that he thought our boss had been playing dirty tricks.

'I'll show you the best way home,' he offered.

I hesitated, then noted the day still young. To hell with Jimmie. Old John would certainly get a visit.

When Alec heard how I planned to go, he suggested he came with me for some of the way. We climbed up the ride, companionably and slowly, through a green and pleasant place with spruce on either side. I asked him about the heronry. It was old, he reported, no one quite knew how old and because it was difficult to find, except from the loch, the birds were seldom disturbed. Perhaps, when the spring came and they were nesting, we could take a look at it together? Thus a seed was sown and a friend discovered. I gladly agreed.

'Jimmie is an interesting chap,' I remarked cautiously a little while later, not wanting to put my foot in it but hoping for some clues about the man.

Alec chuckled. 'He has peculiar ways. You'll get used to him.'

We chatted about the stalker's job in the forest and I learned there was a big deer problem. Every winter, heavy falls of snow on the hill built up against the fences and became hard-packed and solid. The deer used these convenient platforms as a means of leaping over into the newly fenced plantations. Alec was kept busy either driving them out or culling them to reduce their numbers to acceptable levels.

'I saw a pair of eagles this morning,' I reported. 'Is there an eyrie in the forest?'

'Oh yes,' he confirmed. 'We have three pairs altogether. Are you interested?'

'Certainly am, and in any other wildlife.'

'There are only one or two roads into the forest as yet, so few people come this way. There's very little disturbance. When you're settled in, you might like to go out with me.'

'That would be great.'

It took nearly an hour to scramble to the top of that long ride. Then we sat on some rocks for a spell, Alec surveying with satisfaction a scene he knew well and I eagerly taking in the immensity and potential of the job I would be doing. Down below, a long way off, a small breeze was driving little wavelets of molten silver over the loch. I could just see the aspen wood with its bulky cargo of nests and binoculars picked out a heron alighting on one of them. This exciting place was going to be my 'kingdom' and I cast a proprietory eye over its wide expanse. Such things as orders from on high did not enter my mind!

'Right,' said Alec, at last. 'I'd better be off. Carry on along the hillside for a couple of miles and you'll come to a larch wood. There's a burn in the middle of it which is Old John's water supply. Follow it down.'

'Thanks, a lot,' I said. 'See you later.'

A young roe buck has antlers in velvet and a winter coat that is moulting.

Roe deer does seek shelter in the larch wood.

Beautifully marked, this roe deer doe blends with her surroundings.

In winter, the red deer hind mostly has a dark brown or grey coat.

Outwith the rutting season, red deer stags are often together in large groups.

"Greetings, friend!"

Who goes there? An unusual stance for an otter on dry land,
but frequently adopted in shallow water when checking for scent.

A dog otter was taking a rest, but he was always alert to a stalking photographer.

A young otter climbs ashore.

Total trust! A vixen, in the wild, would never allow this
close encounter with just any photographer.

The vixen does not allow her mate to enter the den.

What's out there? A cautious young cub was curious.

Mostly creatures of dusk and dawn, this young pine marten
was leaving the den to join its siblings at play.

Its parent tested the air for scent before setting off to hunt.

A stoat was slinking off to a nearby rabbit warren.

The weasel is a formidable predator, as well, but explores
smaller runways and tunnels after voles and mice.

Pole cats were once restricted to Wales and nearby counties, but numbers have increased since persecution ceased. Often confused with pole-cat/ferret crosses.

Though legally protected, the wild cat is a much-persecuted animal and therefore difficult to locate and stalk. This one was hunting across a rocky hillside.

At dusk, badgers come out to feed and play.

Nuts will tempt the red squirrel to a feast.

Hunted by golden eagle and other predators, this mountain hare
was surveying its world from the safety of a snow hole.

From its den among the tree roots on a river bank, this mink was testing the air, perhaps for danger, perhaps for prey.

Though well-camouflaged with its surroundings, the adder
is sometimes the prey of the golden eagle.

From a hide on the hillside, the photographer caught this golden eagle
flying from her eyrie to hunt prey for her chicks.

Still darker than its parents and with enormous feet,
this eaglet was almost ready to fledge.

This buzzard nested on a cliff ledge. She was photographed
flying in with prey for her chick.

On her platform nest in a larch, the sparrow hawk hen
presented a morsel of prey to her chick.

Oh dear! What happened? A young sparrow hawk 'mistake'.

Her eyrie, a bare ledge on a cliff, the peregrine falcon brought prey to her chicks.

Not yet fully fledged, they perched awkwardly on the nearby
rock face and for a while were fed by their parents.

The female hen harrier takes a rest from brooding duties
in a sapling pine close to her nest.

She flies down to her nest in the heather.

It was slow going over the rough, heather-clad hillside. There were burns to cross, scree and boulders from high on the hill to negotiate, and a fair amount of peat bog, as well. It was nearly an hour before I came to tall and gracious larches on a steep slope below me, planted some time ago and doing well. I climbed a small bluff to get a better view of the shore. And there it was. A small grey house beside the loch, seeming to grow right out of the stones and boulders of a little bay. Mark without a doubt. A wisp of blue woodsmoke swirled around its chimney stack and wafted indolently over the water. Behind was a small field, a sheep fank in it but no sign of the errant sheep Jimmie had mentioned. Perhaps they were hidden in the birch and oak which lined the shore. A plantation of young spruce spread up the hillside. John's ewes were probably there helping themselves to the tender saplings. I suddenly felt very new to the job. How did one tackle the old man on this thorny subject?

As Alec had promised, a gully split the larch wood in half. Within it a burn gurgled merrily along and was, presumably, the aforesaid water supply. I set off down the hill beside it, easy walking in and out of tall trees, a path over larch needles on a relatively gentle gradient. It was a pretty place, sunshine and shade, gently stirring branches in the canopy, an almost incandescent light reflecting the gold of autumn needles. Moss-covered rocks lined the burn and in spring there would be primroses, violets and bluebells growing on its banks. I kept an eye open for small, untidy, platforms of sticks high among the larch branches – I knew sparrow hawks often nested in these trees.

Dreaming a bit and thinking of a break for lunch, I came to the top of a small crag, its sheer face falling away below into awkward rocks and giant boulders which continued on, a hundred feet or more to the bottom. Now I was looking down on the tops of larches, a restless sea of pale gold branches swaying in the breeze. The burn had become an impressive little waterfall tumbling over the hillside and hurrying to the loch. No path that way. I wandered along the top looking for a route and soon discovered a narrow deer track zig-zagging through rocks and heather on a less steep gradient. It held a surprise.

In a sun-dappled clearing on one side of it lay six huge black cattle beasts, magnificent animals, each with a thick shaggy coat. Massive jaws revolved in steady rhythm as they chewed the cud. They looked thoroughly at home as if this was their accustomed 'place'. 1 hastily withdrew a few feet, lay on my stomach and cautiously wriggled forward again hoping I was invisible. Where had they come from? They certainly had no business in the forest and looked ridiculously out of place. Binoculars gave me a better view. Their rough coats were filthy and matted but they appeared well enough fed. All had impressive heads of strangely curving horns. Two were bulls, so this made it unlikely they were someone's domestic animals gone astray; no farmer would risk putting two males together in such a small group. Could these be truly wild cattle still existing in a place seldom visited by humans?

One of the cows glanced up. Still chewing contentedly, she regarded her companions, all similarly occupied, then rubbed her nose against a foreleg. Suddenly, she hesitated, lifted her head and stopped chewing. She seemed to be looking in my direction. What had startled her? Could she have scented me? Her eyes passed slowly along the cliff to the point where I was crouching, paused, then discovered me. For a second she was still and then, a silent message not noted by me, five pairs of startled eyes had joined hers. She bellowed and a bull echoed her warning. All six struggled hurriedly, clumsily, to their feet, turned on the spot, kicked up their heels, and galloped away into the wood. For several seconds the sound of their crashing, splintering hooves echoed through the trees. Then there was silence.

I ate my lunch, picked a way down the deer track and returned to the gully which would take me to Mark. The larch wood came to an end. Norway spruce took its place, a young plantation stretching away to my left. An invigorating smell of seaweed told me the shore must be close. Old John's house could not be far away. I came to an old oak wood which had been under-planted with spruce. A well-worn track led through it. It was the custom then to plant existing deciduous woodland with spruce and often to ring the oak, or other trees as well, so that they died and eventually let in light to encourage the healthy growth of the conifers. But here, mixed incongruously with old trees already ringed were young spruce saplings looking anything but well. It suddenly dawned on me that where there should be no sheep there were certainly sheep droppings.

The evidence was damning. Most of the small trees had been cropped right down to the ground. Wandering slowly along, noting more and more of the poor misshapen things, I soon discovered the cause. A black-faced ewe, well-fed and plump, was quietly browsing on the lower branches of one of the trees. She nibbled daintily, chewed busily and then lifted an enquiring nose to seek out more. I shooed her away but she was not impressed, walking unhurriedly only a few yards to another. Then I passed groups of three or four, all obviously finding the feeding good. By the time I had reached the end of the wood I had counted thirty animals. Presumably they were Old John's. No wonder Jimmie thought a bit of supervision was necessary.

A broad track led straight to the little house. Like my own at Coillessan, it looked old and rundown, its roof none too secure and the stones of its walls crumbling. An old garden wall surrounded it, overgrown with brambles and patched with fencing wire. The building stood close to the shore in a small bay pebble-beached and sheltered at either end by a small spit of pine-topped rock. Wind and tide rolled small white horses over it and a green dinghy, seemingly on its way to the outer loch, was making heavy weather against them. A man was pulling hard at the oars and in the stern was a bundle which looked remarkably like a small Christmas tree. Another man, whom I took to be Old John, stood on the shore watching. The assumption must be that he had cut one of the estate's trees and given it to someone from over the water. What other mischief was this old reprobate up to, I wondered?

I started across the shore and the dogs heard me coming. One by one they leapt to their feet and began barking. A rough-haired hooligan with an impressive white ruff led away, and then, almost as one, they were tearing towards me. Pandemonium let loose came racing over the pebbles and soon, with tails wagging furiously, was leaping all over me. It was a typical exuberant collie welcome. Their owner came hobbling after whistling and shouting in vain. Puffing and out of breath he laid about them with an old blackthorn stick and roundly cursed his unruly mob in Gaelic. One by one they settled at his feet.

'They don't see many people,' he apologised profusely, and turned to subdue an excitable pup.

'I'm Don MacCaskill, the new forester,' I explained. 'Jimmie Reid sent me down to see you.'

He shook my hand but did not seem especially enthusiastic. 'They always come by boat.' he added, as if I had broken the rules.

'I met the stalker in the forest. He told me. The launch is having an overhaul.'

He made no comment but stood gazing out over the loch again. I took a surreptitious look at the man. Unshaven, with at least a week's grey growth on chin and cheeks, an ancient checked cap drooped sadly over his forehead and the dark, once-blue Guernsey and brown tweed trousers were filthy. There were holes in the legs of his wellington boots, as if the puppy had been at work.

'How are you?' I asked, at last, when the silence had grown long.

'Fine,' he replied, then spoke sharply to restless dogs.

'Lovely place you have down here,' I tried again.

'Aye.'

We were not making progress. Better get to the point. 'Are those your sheep, John?'

Startling blue eyes met mine without a blink. 'Aye.'

'They're damaging the trees.'

'Jimmie Reid says I can have them here.'

Old John evaded actual admission of guilt and took all of half-an-hour to do it. The damage to the trees. Was there any?

There was, indeed.

Strange, he'd never noticed it. Pause. What could be the cause? Did I really think it was the sheep?

I did.

Pause.

They must be getting out through holes in the fence, then, though he'd never seen it happen.

Perhaps, but why were the fences not mended?

Long pause while he considered this one.

Well, Jimmie had not sent any netting.

My eyes went to a bundle of old and rusting fence wire against the wall of the garden.

There's been no time ... his eyes had followed mine.

I would send him some more as soon as the launch was ready.

It would be a while before he could attend to it. The new planting kept him busy.

I gave up on that one and changed the subject. The man pulling out of the bay – an early tree for Christmas?

A friend had been over to collect one.

I quoted the rules of the forest, that the sale of all trees had to be through the Office.

He was quite unabashed. Was that so? He'd never heard of such a rule. It had just been a wee present for the kiddies.

Anxious to establish some kind of rapport with the old man, I switched to my adventure of the morning.

'I came across cattle in the forest, on my way down the hill. Seemed pretty wild. What are they doing there? Whose are they?'

The relief on the wrinkled face was comical. 'They're Mister Blair's from the farm on the other side of the ridge,' he readily volunteered.

'Why are they over here?'

'The high ground is unplanted. No fences. They wander all over the place.'

'Has this been going on long?'

'A few years. Each lot of calves is wilder than the last.' John was warming to his tale. 'Maybe, they're doing the damage to the trees.'

'I don't think so, John.' I smiled.

Suddenly, I remembered the short November day. The sun was already well into the west.

'I must be going,' I said. 'It'll be dark soon and it's a long haul up the hill. By the way, how does your pay come to you?'

'I cross the loch to Finnan,' he replied, as if it was the most trivial of undertakings. 'I pick up the train to Tarbet, then walk to the Office. Angus gives me my orders there.'

I thought he was kidding. The man was small, spare, maybe close to retirement, and looked frail.

'You cross the loch?' I repeated, in astonishment, glancing over at least a mile of restless water.

'Aye,' he acknowledged, but added no more.

'Well,' I said, needing to mull that one over. 'Cheerio for now. I'll be back with that fencing material in a day or two.'

He made no comment, but instead, offered some advice. 'You can go back by the shore now the tide's going out. It's shorter that way but you'll need to climb the rocks at Rhuveag. Watch out for the cattle beasts. They're dangerous.'

If Old John could manage along the shore, so could I. I set off across the bay, first over pebbles which crunched and slid away with a clatter, then on to short-cropped grass sprinkled with sand and seaweed – big seas in winter, I thought. On an impulse, I turned to the lonely figure still watching me go. His dogs were romping all about him and he raised his stick in salute. I resolved to ask for regular use of the launch, so that an eye could be kept on the old chap ... and his sheep.

By now the sun had disappeared from the west side of the loch. The forest above the birches on my left was shadowed and promising the darkness of night. High on the hill on the other side, golden light was still blazing the bracken to a fiery rust. Nice contrast. The wind had freshened. Waves were crashing against the encircling arms of the bay. I wondered how the owner of an illicit Christmas tree was getting on. Better get a move on, though.

At the far end of John's bay there were rocks to climb. No problem. It was when I was clambering down their far side that a peculiar boulder on the shore caught my attention. It was about a hundred yards away and in the dusk looked black, possibly volcanic in origin, an odd one out in the surrounding rockscape. Driftwood, seaweed, and dead bleached branches all washed in by the tide, had built up around it and over its top. At springs, it would be covered. Worth taking a look, I thought.

As I walked towards the strange rock, it suddenly began heaving up and down and the branches over its top shaking crazily every which way. And yet there was no wind to speak of. Something made me approach cautiously. As well. One of the wild cattle I had seen that morning lay jammed solid between two rocks. A bull! His enormous bulk was my rock and his impressive horns the strangely waving branches above. How the poor beast had got there was a mystery, for his legs were free but his shoulders held fast. He lay panting, the foam slavering from his mouth. When he saw me he went mad. Wild eyes reflected fear then fury. Then he began frantically kicking and wriggling his huge body, trying everything he knew to free his imprisoned shoulders. In vain.

What the hell to do? How to shift those rocks? How to avoid those flaying hooves? I looked hurriedly around. At least the tide was ebbing and he was not in danger of drowning. A piece of driftwood was handy, a long spar which looked solid enough. I dragged it from a mass of debris then approached the animal cautiously. A deafening roar and renewed thrashing almost sent me flying. This is impossible, I thought. But the spar was long enough. Better have a go. Murmuring soothing words to the stricken creature, I dug in beneath one of the offending rocks and, using the spar as a lever, put all my weight on it. There was no movement at all. Again and again, I tried. It was no use. The rock was unyielding and I cursed the silly beast as he tossed his well-armed head and struck out with his formidable feet. The idiot was ordering his own funeral – the night's tide would be high and he would drown.

I paused for a breather, noted my legs shaking and my heart thumping uncomfortably. I'm running out of steam, I thought. The poor thing's had it. One more shot, and that's it. I put all my weight into a gigantic heave and perhaps, in desperation, a bit more. It did the trick. The rock shifted, came up with a most satisfactory 'glug', then rolled over to one side. There was an explosion of released sand and small pebbles, and then the panicked beast was struggling to his feet. He was free.

Neither he nor I could believe it and for just a moment we stood with heaving sides glaring at one another. I caught a certain look in his eyes, calculating and malevolent, and fled for the rocks on the outcrop. I threw myself up the nearest rock face and, grabbing a small rowan branch above my head, pulled myself on to a narrow ledge. Just in time. The maddened creature came charging in. Crash, his great horns hit the rock below and with a horrible jarring and clattering scraped back down again. There was gratitude for you!

I glanced up quickly for a safer ledge. There was none and the smooth rock looked un-climbable to the top. I turned my head to check on the beast. Head lowered and tail lashing, he was galloping in once more. He slammed into the rock below and grunted with rage as he hit the unyielding surface. I wished my feet were smaller! Recovering yet again, he began to paw the ground, all the while looking up at me, eyes rolling wildly and snorting angrily. What was wrong with the silly beast? He had his freedom. Why did he not go? In the heat of the moment and near panic, too, I could only think he held me responsible for his recent predicament and wanted revenge.

Now, the crazy brute tried a new tack. He trotted back a few yards, turned in a tight circle then, all evil intention and malevolence, came scorching in towards me again. With an astonishing leap worthy of a Grand National winner, he took off. Contact. One of his horns caught my boot. I yelled, grabbed the rowan branch for safety and let fly with the other. I caught him right between the eyes. He grunted and fell back to the ground with a terrible thump. At last he'd had enough. He staggered to his feet, rose on hind legs to pivot round, then roaring with rage, belted for the forest. Never a dull moment in this place I thought, as the sound of crashing hooves faded into the distance and I slid down from my ledge. I wondered where the rest of the wild cattle were and made a note to speak to Alec. These animals certainly were dangerous.

The rest of the journey home was uneventful and shorter than in the morning. The shore was safe, though sometimes slow, and if I had persevered before, tide or no, I would have made it over the rocks of each promontory. As the moon rose over the loch, lighting up still water with cold, incandescent brilliance, I felt a certain satisfaction that I could report to Jimmy Reid a successful foray into the forest. Three hours later, tired but elated, I walked through the Wood of the Waterfalls to my new home.

four: old woodsman

THE mask on Jimmie's face slipped a little when I told him I had been to Mark.

'I see,' he acknowledged, 'You, er, walked down?'

'Yes, I did. The stalker tells me I should have gone in the launch. When will it be available? '

Jimmie seemed lost for words.

'We need to do something about Old John,' I continued, sensing a certain something in the man. Surprise? Discomfiture? Amusement? 'He needs material to repair the fence round his field. Those trees are in a bad way.'

'Excuse me,' my boss hurriedly interjected, then took himself out of the room. I heard a familiar creak of hinges on a little cupboard door, a pause, hinges again, and then he was back.

'Right,' he said, obviously now restored and able to face the day. 'The launch will not be ready until the beginning of January – a new engine is being fitted. John'll have to wait. I've no work for you 'til then. Get to know the forest. Take some leave if you like.'

Once again I left the office wondering what kind of a Head Forester was this? Why send me off to Mark without a map of the forest? It was a hell of a long way and a route difficult to find. Did he think I needed taking down a peg or two? And then, in respect of the immediate future, surely there was some work he could give me, just out of college or not? Excessively shy? Just surly? Was there a drink problem? I hoped Alec was right and I would eventually get used to his strange ways.

Leave seemed an awful waste of time. I didn't take it. Instead I explored the forest and found it impressive. Planted on Scandinavian lines many years ago, there was Scots pine, European larch, Douglas fir and Norway spruce. Some of the woods were quite old with trees at least seventy feet tall. There was medium term forest with trees around twenty five to thirty years old and then the newest section experienced on my way to Old John at Mark which extended way beyond there to the west. Plenty going on: old woods to be managed, woods to be thinned, new woods to be planned and planted. At least I could see why Jimmie had no men to spare.

I made several visits to Arrochar. Surrounded by high peaks the village seemed a gloomy place, the sunshine, when sunshine there was, only briefly warming its old buildings in the short winter days. A vast array of flotsam and jetsam, brought in by the tides or blown in by the prevailing westerlies, perpetually decorated the seaweed shore at the top of the loch. Houses straggled along the shoreline, some stone-built and quite old, others more modern brick bungalows. Smoke drifted from their chimneys in lazy streams and hung, on still days, like a malevolent canopy over all. A small store, which also contained the Post Office, sold nearly everything you could need for day to day living – a long trip to Glasgow by train or bus was necessary for major items and few people owned cars. Once a week a grocery van visited most of the community so, six miles of pedal work away for me, I seldom bothered with the ride. Socialising, when not just dropping in on a friend, took place in either the village hall or the old hotel. Films were sometimes shown in the former, but it was mainly used for ceilidhs and dances, both of which would continue with unabated enthusiasm into the early hours. Coffee mornings, usually organised by the minister's wife to raise money for good causes, and sales of work took place there as well. The bar in the hotel was where serious drinking took place, plots were plotted and gossip considered and discussed. The Forest Office, surrounded by old pine and larch on the edge of the forest, was situated midway between

the last of the village houses and my own down the lonely track to
Coillessan.

An absorbing hobby kept me busy during this time, and was to be
followed up over many years. Dr Neil Crystal, one of the professors at
the College in the Forest of Dean, had inspired an interest in entomology
as it related to woods and forests. He was a well-known authority and an
author as well. Now, I spent long hours around Coillessan turning over
stones, peering under boulders, prodding the vegetation, microscoping
the bark on the trees, stirring up leaves and litter, looking for whatever I
could find. I became fascinated by the many different species present and
collected them, gassed them, and pinned them on to sheets of card.
Later, I came to realise that this was not the best way to learn about these
fascinating creatures and instead, making use of knowledge already
acquired, spent long hours just nosing around and watching. Spiders also
especially interested me and I became positively obsessional about not
disturbing them in their filigree webs.

Signs of wildlife were everywhere. Prints in soil or sand, droppings on
small rocks or on narrow tracks threading through the trees, all told a
story – the stalker was certainly right about there being too many deer –
and old nests high in the trees suggested buzzard, sparrowhawk, jay and
hoodie crow nesting in the spring. I heard a pair of ravens 'croaking'
high over the ridge above the forest and thought again of the eagles I had
seen on the way to Mark. A red squirrel, out on a rare warm day, gave
away his home by leaping, branch by branch, towards a beautifully
constructed drey in the fork of a larch tree. No red squirrels at home in
Kilmartin, so a bonus. I also spent a few dawns and dusks hidden behind
a rock close to the shore hoping an otter would cross on its way to the
loch. It never did though I found fresh droppings. Many years later, I
learned what a sensitive nose the animal has. Perhaps my scent was
everywhere.

At last, shortly into the New Year, I was called to the Office. I found
Jimmie, wig in place, immaculately dressed, and in full command of
himself. In fact he was positively affable.

'I've good news for you.' He came straight to the point. 'The Douglas
wood is to be thinned and the Norway plantation to the west of it must

be brashed. The squads will report to you tomorrow.'

'Great,' I replied, surprised but suddenly nervous – I'd hardly met any of them. 'Who are the gangers?'

'Andy Murray and Bill O'Connor. Good chaps.'

'Right,' I said, hoping I sounded confident. 'I'll get up to the Douglas today and mark the trees.'

'Very good,' acknowledged my inspiring boss and, true to form, immediately took refuge in the pile of papers on his desk. The interview was over.

A job to do, at last. I cycled home, elated but afraid I'd make a bloody fool of myself the next day. Having arrived at the end of the track, it was automatic to take a look over forest and loch, the one beckoning – there is work for you now – the other sparkling in the first sunshine for weeks. A surprise. The *Mary Ann*. Jimmie had said nothing but there she was in the bay, all forty feet of her, in fresh paint and shining varnish. With binoculars, I took a look. No cockpit. Seating built along her sides. Plenty of room midships for materials and tools. An inboard engine, 5 knots I reckoned, but no more would be needed. She proudly breasted each choppy wavelet stimulating, in me, immediate thought for the next visit to Old John down at Mark.

The second surprise was in my yard. Alec, the stalker, was crossing the cobblestones from the stable brushing dust and straw from his breeches.

'Hello,' he greeted me. 'Happy New Year. You've got company.' I had, indeed. Restless hooves on the stable floor and a whinny meant ponies on board.

'They're from Ardoch,' the stalker explained. 'Both garrons. Used to working in the forest. The horsemen are still on holiday. Jimmie asked me to walk them down. This is Jock and that's Tommie.'

'Good to see you,' I managed a word in, at last. 'Happy New Year.'

'Peter and Andy will be out tomorrow.'

'O ... right.'

Alec must have sensed a certain hesitation. 'Didn't Jimmie tell you the ponies were here?' he asked, grinning.

'No, and I haven't met either of the horsemen yet.'

'They're good chaps. Don't worry.'

'As you said, our boss has peculiar ways!' I smiled. 'He didn't mention the launch, either.'

'That's him, all right. You wait till he gets his telescope out!'

'His what?'

'He carries it everywhere, to keep an eye on us all. From the road.'

'Oh sure!' I mocked.

'No, really. It's the climbers. They walk through the forest to The Cobbler and light fires in the caves up there. He's afraid they'll start a big one.'

I still thought he was kidding, although by now nothing that Jimmie did should surprise me. Instead, I began stroking the velvet noses of the two sturdy, rough-coated ponies. The first, called Tommie, was a pale, chocolate coloured roan with a touch of white on his muzzle. The other, Jock, was dark, dappled grey and handsome. I noted their broad shoulders, short thick bodies and powerful quarters, ideal for the work they must do on our steep hillsides.

'Nice beasts,' I commented. 'It's good to have the stables used again.'

That afternoon, I walked quickly up the steep hillside to look at the Douglas Wood and to mark those trees that would be felled. Here was awe and a dreaming of the future, for though a great part of my training meant thinking of all conifers as crops to be grown, harvested and sold for many different uses, I could never forget the sheer miracle of their growth. The Douglas Fir is a particularly impressive species. A tiny seedling, only a quarter of an inch in length, could, given the right conditions, end up a giant maybe 300 feet in height with a base of 15 feet and a possible life span of more than 300 years. I had read much of the man who had given *Pseudosuga dougiassi* his name. Born a poor gardener's son in Scone, Perthshire, David Douglas had become a famous botanist, travelling the world and enduring much hardship in order to discover new species. Amongst his trophies were the seeds of this fir, brought back from British Columbia, and from that seed these trees were descended. As I wandered among them, selecting those that would be cut during the following days, I dreamed that here, too, could be a magnificent wood which many would come to see and admire. About thirty years old, many were of exceptional promise and my job was to see that they received more vital sunshine and light.

The next morning, I awoke to the sound of harness jingling and someone calling words of encouragement to the ponies. The horsemen! Of course, they would be here early to groom and feed their charges. I ate a hurried breakfast, ears flapping to the sound of people arriving on bikes, on foot, and in what sounded like a creaky old banger. There seemed an awful lot of them. I heard cheerful greetings, felt anything but cheerful, and tried to get my act together. During the best part of two months, I had met only one or two of the workforce. How would I get on with them? Would they give a new recruit a hard time? The moment I had trained for had arrived and now I rather wished it would go away. There was a knock on the door and I took a deep breath.

'I'm Andy,' said a lean, tall man as I opened it. I noted steady eyes and mild interest. 'We're all here.'

They sounded like it! Through the open door, I could hear a hum of lively gossip and stamping feet against the cold. A ringing laugh lifted clear above the restless sound, rose into the still air and echoed away over the nearby loch.

'Thanks,' I replied, tongue-tied. And stepped into the yard.

There was an instant silence which seemed to throb with unspoken curiosity. A hell of a lot of them, I thought, though I knew there would only be eighteen. A snorting, blowing and clip-clopping on the cobbles came from the stable door, ponies impatient to be off and working. A horseman held each unruly head. In my panic, I noted only many faces and the cold air above them filled with steaming breath. A few friendly words of introduction? All that came out was: 'Morning all. I'm the new forester. I'm sure we can work together.'

Everyone seemed happy enough with that and we all trooped out of the yard. Andy introduced me to Peter and Dick, the horsemen and Bill O'Connor who was in charge of the girls' squad. The rest I would get to know later. We set off into crisp, frosty air, cheeks smarting, ears tingling, noses dripping, and made for the track which led through the Wood of the Waterfalls. Now I could take a better look at the gang. Dress for the day seemed to be as fancy took each one, but the men were more uniform – thick, dark sweaters, working trousers and, without exception, cloth-caps on their heads. All wore tackety boots. In contrast, the women were more adventurous with gaily coloured sweaters and head scarves, workmanlike slacks or Land Girl breeches and heavy wellington boots. In all, a motley crew. On wet days I was to find everybody in regulation uniform, black oilskins and peaked sou'westers, looking for all the world like busy carrion crows.

As we climbed the steep hillside, through winter-bare oak and birch, steam rose in twin streams from the nostrils of the ponies and floated away through the stricken branches. They trudged up the track ahead of us, a steady, plodding footfall: this is the start of just another day's work, at the end of just another forest track. The rest of us huffed and puffed up the hill, anxious to get warm by moving as fast as possible. All around was cheerful chatter, the subdued swish of cloth against cloth on jackets and trousers, and the rustle of boots through litter on the woodland floor. Peter and Dick kept talking to their charges, patting them, running their hands over sturdy, smooth backs, checking bits carefully to see they were comfortable, and evidently continuing a bonding process begun in the stables. All of these, good familiar, sounds. It was a great day and I began to feel better.

'They're good with the ponies,' I remarked of the horsemen to Andy.

'Oh, aye,' he agreed. 'They've been doing the job a long time. Doesn't matter which ponies, as long as it's ponies!'

O'Connor chipped in. 'Tommie's sweating. He's a bit fresh.' His words were prophetic.

In due course, we reached the main forest road which bisected the planted area of the hillside. On it stood the neat mechanical cross-cutting bench on which shorter lengths were cut into pit props, anything from three to six feet in length and between three inches and six inches in diameter. A 'peeling' spade, a flat, sharp blade on the end of a short shaft, was lying close by. It would be used to remove the bark before sawing. Andy detailed off three men here. Now the squads divided. Bill O'Connor set off with the women for the spruce plantation; I would check on progress there later. The rest of the party began climbing the next three hundred yards, a steep, rough, rock-strewn unplanted area, towards part of the forest where they would be working. Suddenly, I noticed a steel wire overhead. It was rigged between two trees, from an old pine on the edge of the forest to a single Douglas fir which had been left uncut down beside the road. It had been drawn as taut as would be possible. Entering the wood from another point yesterday, I hadn't seen it.

'What's that for?' I asked the ganger, suspicious of unofficial bad practice.

'We were supposed to begin thinning the Douglas back in November. The ponies had not arrived. Jimmie Reid turned a blind eye so that we could get started.'

Did he indeed, I thought! 'I see,' I acknowledged, but did not. 'How did it work?'

'You see those hooks at the bottom end? They were yanked up to the top, a log attached to each, then we gave them a hefty shove.'

'Jesus!' I exclaimed. 'That's dangerous.'

'Aye. They fly off sometimes when the tension's too great. Could land anywhere.'

'I see,' I said again, horrified that such risks had been taken. Ingenious, or not, it would have to be dismantled as soon as possible in case someone thought it would be fun to experiment again. Should I get the men to do it today, or ought I to speak to Jimmie first? It was the first little problem of the day, but I settled for the original programme. A mistake. We reached the Douglas Wood, at last, and I told Andy to carry on. Four of the men would be felling and snedding and the horsemen would load up the logs and lead the ponies down the hill to the road.

For the first time I noticed a hardy old character who, despite the bitter temperature, was busy throwing off his jacket and then the old sweater beneath it. He was short, thickset and, I suspected, immensely strong. Busy rolling up his sleeves, he stood gazing up at the trees with a professional air, as if assessing their wellbeing and potential.

'Who's that?' I asked the ganger quietly, as the old man turned to speak to his mate.

'It's old Dougal McPhail. He's a great old chap and keeps us all in order! Not much he doesn't know about trees.'

I resolved to have a word later.

The Douglas Wood looked bright and welcoming in the winter sunshine and I began to look forward to the day. In a short time, Andy had his squad at work and the sweet rasp of cross-cut saws, one-two, one-two, reverberated through the trees. I watched him and his mate for a while – two men bent almost double, feet firmly anchored, working in perfect rhythm. As the cross-cut saw bit ever deeper, long streamers of sawdust, fine and bleached, showered down to settle about their feet. Then there was silence as each rose to mop his brow. They stood a moment, considering, speaking a word or two. Andy nodded his head, picked up his axe then with care inserted a couple of wedges into the cut. Two blows with the butt, exactly to the point required, was the final prompting. The men stood back and the tree began its fall. The sound of protesting fibre exploded into the expectant hush and branches whipped through neighbouring branches to tear, brush, and bend. With a dull thud and a last great shudder the tree came to rest exactly on the spot intended. The other man, whose name I did not yet know, with rhythmic swing of arm and wrist, began working along the stem with an axe, neatly snedding the branches away. Then it was time for one of the ponies.

I had watched as they were being harnessed up. Tommie had tossed his head, pawed the ground impatiently and kept glancing round at Peter as if unable to understand the delay – a prima donna about to go on stage. Jock, on the other hand, had stood motionless as side chains, swingle tree and chain were fitted. To him a new forest and a new handler were evidently all in the day's work. Now, they stood beside the dragging tackle, breath steaming, ears twitching, long, thick tails swishing lazily.

'What do you think?' I asked Peter.

'He's just fresh,' he assured me again, giving Tommie a pat on the neck and smoothing his thick mane. 'Be all right once we're away.'

'Jock's got more sense,' chuckled Dick as he put the finishing touches to the job. 'Okay. Let's see how they go.'

He led his pony to the tree just felled and Peter took Tommie, his tackle banging and bouncing over the uneven ground, towards the second team a small distance away. The drill is as follows. The pony is led towards the fallen tree, turned around to face the way he will be pulling out, and then backed in to the right position for loading. The horseman attaches a whole stem to the chain, then leads the pony forward so that another can be added. If the tree is too bulky, it will be cut into sections and harnessed up as logs. The load is complete when the horseman judges more would be too great a weight to drag. The animal is then led down the hillside to the forest road where the load is detached.

I watched a perfect performance from Jock. Backed into position, he stood quietly by until the tree, bare of its branches, was secured. Then he moved forward, without an order, for the next. When they were ready to go the pony, with a quiet word from Dick, set off sedately through the trees. The horseman, apparently sensing both experience and reliability, walked quietly along behind letting the animal have his head at the end of the long reins. All went well until they were halfway down the hillside. I saw the pony slow down, then with heels dug in and shoulders taking the strain, come to a halt. The logs slithered awkwardly to rest but did not slide dangerously forward. What was wrong? Dick patted the animal on the neck, rubbed his nose reassuringly, then appeared to steer him round some obstacle or other. Binoculars revealed a large pile of boulders over which it would have been impossible and dangerous to try and drag the load. Then they continued. Dick told me, later, that he'd given no orders. Jock had known best what to do.

Tommie, on the other hand, was throwing his head about fretfully and shifting his weight from one foot to another. He was anxious to be off.

Only one small log in place and the pony was raring to go. I saw Peter move to calm him and sensed trouble.

'I'll give you a hand,' I shouted, and started running.

'He saw Jock away,' Peter called. 'Wanted to follow.'

Suddenly, with a loud whinny, Tommie put his shoulders to the derisory weight behind him, and took off.

'Steady up, Tommie,' we yelled. To no avail. The pony tore away through the trees, the log bouncing perilously from side to side as he dodged around them. He did not draw up at the edge of the wood. Instead, with only a small pause to look about – I think he was trying to find Jock – he bolted on down the hill towards the road.

By now, the silly animal had a fascinated and horrified audience. We watched him careering over that rough hillside, leaping all obstacles, heather, boulders, scree, mane and tail flying, legs lifting, the log bouncing along behind and threatening all the while to tangle and trip. He seemed to have a charmed life. Peter charged down the hillside after him, not quite so spectacularly but certainly as athletic – it would have been funny had it not been serious – but really there was little he could do. Tommie was long past hearing orders and was probably panicking as well.

Halfway down there was a small level piece of ground which fell away almost sheer on the far side, to the slope beyond. The pony reached it safely but as he speeded on, over the edge and out of control, the log became somehow jammed. Impetus kept the animal going forward and, of course, he was suddenly pulled up with a vicious jerk. He crashed down the steep bank on to his side, and the collar, to which the side chains were attached, slid up his neck to his throat. We saw him thrashing about, desperately trying to get to his feet, but the more he struggled the tighter the noose would become. Tommie was in terrible trouble.

'The bugger'll choke. He'll kill himself,' shouted Peter from halfway down. 'I think the tackle's caught in a root.'

'Bring an axe,' I yelled to Andy. 'Get moving, boys.' And we, too, set off down the hill.

As we drew nearer the distressed animal, I wondered whether he had broken a leg so awkwardly did he lie. That would be disaster and I wondered how quickly I could get hold of the stalker to put him out of his misery. Wild eyes met ours as we arrived. The pony was gasping for breath, his huge jaw wide open, tongue and lips all covered in saliva. The log was caught by a thick, straggling root.

'Take care,' I spoke to the men. 'Hurry, Andy.'

The ganger, avoiding convulsive hooves with difficulty, expertly chopped the log in half with a few quick strokes. Tommie was free. The cruel pressure was off. He lay quite still with his eyes closed and now and again a great tremor passed over his frame. But at least he was breathing. As we anxiously watched, the frenzied gasping began to slow, the mouth closed, and in a few minutes breathing had slowed to normal. Suddenly, the ears were twitching. Then Tommie opened his eyes and was taking an interest in life once more. As soon as possible, and as gently as we could, we encouraged him to his feet. Murmuring soft words of comfort, the horseman ran careful, exploring hands over the trembling body.

'Don't think there's any damage done,' he reported, at last. 'Can't believe it.'

'Good work, Andy,' I said. He smiled, I thought with genuine pleasure – he'd not made a mess of a tricky job.

I found myself shaking, too. This could have been an horrific accident to report to Jimmie Reid and this, my first day at work. I gave the pony a few more minutes to recover and then decided it would probably be best for him if he went on working.

'Andy, get back up there and carry on,' I said. 'I'll be along later. Take him down to the road, Peter, and unload as if nothing has happened. Then start him off again.'

'Aye,' Peter replied. 'He'll have learned a lesson.'

And, he had. On his return to the wood Tommie appeared to be a sober pony and a model of good behaviour. From then on, however, we gave him a gallop along the forest road each morning before he was hitched up for work. That way he got rid of some of his high spirits.

Now I walked along to the spruce plantation wondering if there would be more problems there. I had never before worked alongside women, though I knew they had been employed in the forest in large numbers during the war years. It just had not happened. So, once again, the nerves. How did one relate to them? Just as one would to the men? Should allowances be made for feminine frailty?

Not too much, it seemed. As I walked into the wood the sound of singing greeted me, a chorus from a popular musical. Tone deaf, some of those singers! It did not matter. The rhythm was strong and well-marked, matching the sound of regular sawing: brish-brash, brish-brash. I found twelve women of varying ages and builds, slim ones and fat, jolly ones and serious, all busily brashing away in pairs. This is not an easy or particularly pleasant job and is nearly always done in the winter months when the weather can be foul. The object is to make access to a plantation easier, for whatever purpose, and is done on a regular basis every few years and especially before thinning is about to commence. On selected trees all the branches are removed to a height of about six feet. The operator uses a slightly curved saw, about eighteen inches long, on which the teeth are angled so that they will only cut in one direction. It is fixed to a shaft about six feet long which enables him, or her, to reach to the necessary height.

This squad seemed to be happy enough and coping well. Nevertheless, I approached with caution having not a clue how one addressed a gang of twelve women doing what I thought of as men's work. I looked hastily around for Bill O'Connor and found him some distance in the trees. He threw down his saw when he spotted me and strode towards me. Fifty-ish, with greying, crisp curling hair and a bristling brindled beard streaked with red, he did not look as if a bunch of females held any terrors for him.

'Morning, Mr MacCaskill!' he greeted me, cheerfully.

'Morning, Bill.' I acknowledged. 'The name's Don. Everything all right?'

'Fine. We'll be finished at the end of the week.'

'Good. Next job's the fire traces. I'd like to get on as soon as possible.' Drying winds in March and April are a forester's nightmare and I already knew that Jimmie was especially nervous about fire.

'Tell them to knock off, now. We'll have a break and I can meet the squad.'

Bill shouted and blew a whistle – some of the squad were further away in the wood – and in ones and twos they came strolling back into the clearing where their 'piece' bags had been stacked. Most were picking spruce needles from their sweaters or removing head scarves to shake them free of the sharp, intrusive needles which would find a cunning way into the clothing. Heads were shaken, too, and long hair, short hair, fair, dark and red, was swung into place and smoothed with fingers free of gloves. All were looking my way with interest.

'This is Katy,' said Bill, introducing me to a slight girl with a mop of red curls and a freckled face. 'She lets me know if any of the women have a problem.'

Frank blue eyes met mine, and held. 'I see,' I replied, inadequately. 'Hello. Nice day.'

The mischief danced in those eyes but she just grinned and remained silent.

The ganger continued remorselessly: this is Mary, Jane, Liz, Morag, Maggie, Catriona, Muriel, Dorothy ... at which moment I gave up.

'Okay, okay,' I protested. 'Stop there, Bill. Nice to meet you all. I'm Don. I'll get to know you all as we go along.'

We sat around on whatever we could find that was dry to drink companionable mugs of tea or coffee. And then the teasing began.

'Is there anyone coming this weekend, Bill?' Katy started it off. 'Don't forget we love you.'

'What d'you mean?' he asked, pretending surprise.

'Come on, Bill. We know all about it.' Another girl laughed. 'The last one went away.'

'Bill is looking for a wife!' Katy turned to me, the devil in her eyes. 'He advertises in the Oban Times.'

I turned for enlightenment to the ganger, but he had quietly beaten a retreat into the wood.

'It's all right, Don.' Another girl chipped in. 'He doesn't mind. His cottage is up Glen Croe. It's dark and lonely up there. He would like to get married but no-one, once they've been there, will have him.'

'Okay, okay.' I said again, smiling but thinking it time to change the subject. 'Time you got busy again, I reckon.'

After another word with O'Connor, who had returned to set the squad to work, I left the women in order to return to the Douglas Wood. I thought of a slip of a girl with red curls doing a man's job with ease and an Irishman who seemed to take merciless teasing in good part from a bunch of cheeky women. I was soon to learn, however, that he was a strict disciplinarian who stood no nonsense.

By now it was close on midday. On the road below the Douglas wood, neat piles of logs were steadily growing. Dick and his pony were just starting up the hill to collect another load and, the day now warm, sweat shone on the animal's shoulders and quarters. Two men, Bob and the Welshman Dai, were already lifting the latest long stem against the whizzing circular saw and guiding it through. Its engine erupted dirty black smoke into the air and its razor-sharp saw whined high-pitched falsetto as it bit into the wood.

'Is Tommie behaving himself?' I asked.

'Can't keep up with him,' they joked.

At the lunch break, we all gathered in a clearing, sunshine on autumn twigs, bleached grass and blackened heather, dappled light on the forest floor. It was good to rest a while and to get to know some of the men. Andy, the ganger, had no problems to report and he confirmed that the pony, Tommie, had settled after his nasty experience and was doing well. A satisfactory number of trees had been felled and the work seemed to be going to schedule. But as we sat, the ponies munching from their bags, the men gossiping, smoking and sipping mugs of steaming tea, I began to sense expectation in the air. Something was up. Something was planned. The men kept looking at Old Dougal who was sitting apart with his back against a trunk, his pipe comfortably smoking. This was the old chap I had noticed in the morning.

'Come on, Dougal,' one of them called, at last. 'Show Don how you do it.'

'He's amazing,' Andy whispered. 'You'd never know he was nearly seventy.'

The old man was obviously pleased, but feigned surprise.

'Come on, Dougal,' I added my voice. 'I'd like to see this.'

With a show of reluctance the man rose to his feet: surely Mr MacCaskill would not want to be bothered with a piece of nonsense?

Oh but I would – I rose to the occasion!

Permission granted and his mind made up, the mood changed. All right, he seemed to say. This is a rare performance, something special. I need your complete attention.

The drama commenced.

Did anyone happen to have a piece of string?

After much delving into pockets, the men all solemnly shook their heads.

Oh, well. Perhaps he had something himself.

Dougal tried each of his trouser pockets. Ah! And he drew out a length slowly, as if, now found, he was anxious that perhaps it would not do. Then, with solemn ceremony, he tied the string around a tree which had been marked for felling. Tested it carefully – no slack could be allowed. At last satisfied and ready to go, he spat on each of his hands and with consummate gamesmanship – this was serious business – picked up his seven pound axe. The edge was examined with screwed-up, calculating eyes and a horny thumb was drawn experimentally across. Then he took a long look round his circle of admiring colleagues. Had he their full attention? The moment had arrived.

With no hesitation or break in the steady rhythm of his arm, the old man worked his way round the stem, first with a downward cut from above, close to the marked line, then straight across its upper edge, until the circle was complete. The string was still in place and intact. The performance was masterly and Old Dougal was obviously pleased with our spontaneous applause. When someone demanded an encore he was quite ready to oblige. Smiling, he strode over to the stump of a tree already felled, and almost faster than our eyes could follow, carved a 'v'cut cross upon it in four clean, unhesitating strokes. A rapid one, two, three, four, and there it was. Perfect.

That was great,' I said, when the applause had died down. And meant it. Axemen often demonstrated their skills in this fashion but I had never seen it done so well, so quickly and with such a professional air. In due course I learned that the forest, and all the happenings within it, was what Dougal's life was mostly about. It was his great interest. Often I would catch him thoughtfully rubbing together a handful of Douglas needles, sniffing their pungent scent with evident enjoyment and gazing up into the canopy with speculative eyes. Maybe he envisaged dappled sunshine and shade in a wood of giants one day. Though now as fit as they come, strong as an ox, broad shouldered and bandy legged, he was nevertheless almost seventy and feared that any day now he would be given notice and forced to retire.

The old man was an almost fanatical gardener, too, and in the spring and summer months was to be found working in his garden long before setting off to join his squad. Then, after a hard day in the forest, he would toil away until bedtime. Sometimes, passersby, who were undetected, would hear him talking to his vegetables encouraging them to grow stronger and taller, better and more beautiful. He was the only man I knew who trimmed a beech hedge with secateurs. Each twig would receive loving attention and the end result was a creation of beauty, thick and solid at the base and gradually building to an even wedge at the top.

Later that afternoon, back at work again, the gentle breeze faded away and sunshine vanished behind a dull grey overcast. It began to be oppressively warm and humid. Great banks of threatening cloud were building in the southern sky and the loch, down below, was utterly still, mirroring their dark menace. I looked around uneasily. Andy was wiping his brow and gazing up at the needled canopy. Old Dougal had stopped work and was donning an old waterproof. The two other men, ignoring the ominous signs, were still snedding. Both ponies had started up the hill with the horsemen. They were restless and each was being led by his halter. No rain as yet but the air was stifling and full of electricity. A storm was surely brewing. At what point should I stop work and tell the squad to pack it in? Would Bill have problems with nervous women?

It all happened too soon. Behind the dense curtain of Douglas tops, their branches swaying to an unseen force, half seen, half sensed, there was an ominous flickering across the leaden sky. Lightening came forking across the rolling cumulus and could be seen dancing in its reflection in the loch. Utter silence. The forest waited. Suddenly, right over the hillside below us, another spectacular zig-zagging through the heavens and, simultaneously, a mighty rumble. The air split apart and there was a sickening thump as lightening struck.

'Take cover,' I yelled to anyone who could hear. I led the way deeper into the wood, then drew-up sharply. The ponies? The horsemen?

'Stay where you are,' I shouted to Andy, and ran back to the edge of trees.

Peter and Dick, with their charges, were still only halfway up the hill. All at once, there was a tremendous crack from above and sinister blue light danced, sizzled, spat down the old highline wire. Another shattering reverberation. It was too much for Tommie. Both ponies whinnied alarm, threw up their heads and reared. Dick held Jock, but Peter fielded a nasty kick from Tommie and was sent flying. Tommie took off. Panicked beyond hearing commands, he galloped off down the rough hillside once again, all the gear behind him threatening disaster. Oh God, what next? By a miracle, he reached the road safely. I saw the men, down there, running alongside for a short distance but they were unable to stop him. Last seen, he was galloping for dear life round a bend.

Then the sullen skies opened. Rain of tropical intensity, solid and relentless, came belting down. Douglas branches bent with the weight. A mosaic of rain-patterned puddles soon covered the forest floor. The drains began to fill. I hurried to rejoin the squad.

'Do you often have storms like this?' I asked Old Dougal who, another ancient waterproof over his head and shoulders, was standing near.

'Never known one like it,' he replied. But wise eyes examined the sky all around. 'It's nearly over, now.'

He was right. The sky was clearing to the south, the rain easing. Short and sharp had been our storm. As we emerged from the wood Dick was just arriving with Jock. The horseman was soaked to the skin but cheerful.

'That's a great pony,' he remarked, giving Jock a pat and a carrot from his pocket. 'He settled at once.'

'Is Peter okay?' I asked, anxiously hoping I was not, after all, having to report a casualty to Jimmie Reid

'Aye. He's got a nasty bruise but he walked down to the road all right.'

There was now no reason to hang about. I sent Andy off to check that Bill and the girls were safe and to tell them to keep an eye open for the errant pony on their way home. I would first check at the stables before sending out a search party.

There was no need. As we trooped into the yard at Coillessan, we found Katy of the women's squad already there and standing at the door to Tommie's stall. The pony, his head resting comfortably over her shoulder, was receiving all the soothing attention he could possibly want.

'He's all right, Don,' called the ex-Land Girl.

'What happened?' I asked.

'We were just packing up to go home when he came crashing through the wood. The trees slowed him down and he came to a halt right beside us.'

I thought it best to lead him home,' O'Connor chipped in, as he emerged from within.

Peter and Dick stayed to groom and feed the ponies and I sent the rest of the workforce home. As they all wandered away, dripping wet scarecrows but cheerful, I thought what a day for a first day on duty and thanked my lucky stars there were no disasters to report. That 'highline' would come down tomorrow.

five: sheeps is cheerier

OVER the next few months, work in the forest was routine for winter time. The women finished brashing in the spruce wood and before they went to another similar job I sent them off with Bill O'Connor to check the fire traces – six foot wide gaps left unplanted between the forest and a public road. This was done well before the prevailing winds in March turned easterly and rainless, which they nearly always did, drying-out the vegetation to a greatly increased risk of fire. Bill's girls would each have used a 'screefer', a tool a bit like a rake but with a blade instead of prongs. This they used to scrape to one side soil with the previous year's growth on it, thus removing material upon which fire could feed. Any worker discovered smoking in the forest would receive short shrift from Jimmie Reid.

January and February were wet, either with snow which soon melted or with the rain, often heavy, which took its place. The forest at its gloomiest. Branches in the canopy dripped perpetual moisture and the forest floor became slippery with mud and puddles, difficult conditions in which to find secure foothold when felling. The burns in the Coillessan Wood all hurried, full, white and foaming, to the loch, each waterfall a spectacular, uncontrollable force sweeping away all that stood in its way. Much higher up, on the tops, the snow was gradually disappearing and unsullied, sparkling splendour had become streaked white upon drab, brooding grey. Encouragement for me, however, was an improvement in relations with the boss. A visit to the office now frequently included a chat about some forest problem or other, or an item of mutual interest. I got the impression that he was beginning to trust my judgment and actually approved of some of the suggestions I hesitatingly put forward. However, one afternoon when he had sent for me specially, he had a proposal which took me completely by surprise. Almost lost in the haze of blue smoke from his pipe, Katy, leader of the women's squad, was sitting demurely in the dingy little office. Jimmie came straight to the point.

'Come in, Don,' he greeted me, affably. 'Katy has come to see me about a new job.'

Why, I wondered? The girl always seemed happy enough and appeared to enjoy her work with the other girls. What was she up to? I glanced her way, looking for a clue, but she said nothing.

'We need to get on with the thinning,' Jimmie explained. 'We're falling behind. I'm taking on another squad and there'll be two new ponies shortly. Katy wants to work one of them.'

'What!' I exclaimed.

'She has had some experience with Clydesdales, as a Land Girl,' he went on, 'the garrons should not be a problem.'

I looked doubtfully at the slight figure. Beneath the mop of red curls, the direct blue eyes held a challenge and a tentative smile played among the freckles on her face. She might well have a way with horses, and men, I thought, but she could never cope with a tough job hauling logs about in the forest. And what about all that heavy tackle?

'I know I can do it, Don.' She spoke at last.

I glanced at Jimmie who was studiously examining a document. He only grunted. I guessed Katy had already won him over.

'Right,' I agreed reluctantly, thinking it was all very well for him to be magnanimous but it would be my responsibility if anything happened to the girl. She could easily be injured by a nasty kick or a falling piece of machinery.

'Where will she be working?'

'I thought in the Douglas Wood, first, till we see how she gets on. Peter and Dick, with Tommie and Jock, will join the new squad at Allt na Fuarain.'

The girl looked as though butter wouldn't melt in her pretty mouth.

'I'll see you out at the stables,' I growled. 'When do the ponies arrive?'

'Day after tomorrow,' the boss informed us. 'Katy should spend a couple of days getting to know them.'

'Thank you, Mr Reid,' said the girl, with a broad smile, and vanished quickly through the door before he could change his mind.

'She'll be okay, Don,' Jimmie reassured me as I rose to follow. 'Don't worry.'

It's all very well for you, mate, I thought, and cycled home mulling over possible problems. How would the men react to having a woman working with them? Would the other women be jealous, thinking she had been singled out for special favours? Could she really do the job?

As promised, the ponies arrived at Coillessan and Katy was there to welcome them. Three-year old mares, chestnut and grey, rough-coated, sturdy and strong, they were called Mollie and Mary. Like Jock and Tommie, they were from another forest and trained to the job. I looked at them doubtfully for they looked enormous beside the petite girl.

'You are to choose which one you like,' said Alec, who had brought them along, 'then Peter will give the pair of you a work out. See if you know your stuff!' he teased.

'Of course, I do,' the girl rejoined, indignantly. But she was obviously thrilled.

'Which one is it to be?' I asked.

The chestnut,' she replied. 'She's gorgeous.'

So, for the next few days, Katy spent most of her time with Mollie. Never could that pony have been so thoroughly groomed or her feeding so full of special titbits. The girl was hardly out of the stable and if she could have slept there, she would have. I made her free of my kitchen and if work took me back there at lunch time, I usually found her having a cup of tea and positively oozing contentment. Peter took both ponies into the forest to try them out and found them docile and easy to handle. He gave the girl a lesson in harnessing her charge, fitting collar, chains and swingle tree. Mollie stood patiently by as they rehearsed, munching oats, turning her head from time to time to see what was going on, apparently unperturbed by her new handler, and a female one at that.

Katy soon mastered the drill. A trial run in the forest was now in order. The whole team set out, Peter, with Katy and Mollie, and the other horseman, a chap call Stewart, with Mary the second garron. I watched them away, over the track through the Coillessan Wood. Two ponies with two horsemen. Only one was a girl.

'Just let me know if there are any problems,' I said to Andy, the ganger. 'I'll leave them to get on with it for a few days.'

In fact it was nearly a week before I set off to see how things were going for our ex-Land Girl. Work with the other squad had taken longer to get organised than expected because the weather had been dreadful. One evening, I met Peter on my way home from the Office.

'Everything all right?' I asked him.

'That girl!' he exclaimed. 'She's fantastic!'

'Okay, Peter,' I said, smiling at his enthusiasm. 'Get back to work with Tommie tomorrow and I'll go up to see how she is managing. Thanks.'

The next morning I pedalled along the road below the Douglas Wood. A breeze from the north and east perhaps heralded change and, at last, though overcast, we had a dry day. Scent and sound pleasing to a forester came wafting down the hillside: pungent smell of resin; rattle of harness; rude rasping of cross-cut saws. This wood was taking shape, now. I could almost imagine the great man himself, David Douglas, approving its potential, and each tree welcoming both light and space in which to stretch its limbs and grow. Once again, I was well-content with my job. On the roadside, huge piles of neatly stacked logs of varying lengths awaited the arrival of the timber lorry. A team was at work peeling and cross-cutting each load as it came. Stewart and Mary were just starting up the hill but there was no sign of Katy. Not wanting to appear to be checking-up, I decided to wait a while in the shadow of some logs until she came down. I would speak with her then.

My binoculars, however, were too great a temptation. I had a quick look. By now the two tracks up the hillside were rough and rutted but mostly clear of vegetation from the continuous plodding of pony feet and the scraping of the logs. At the top of one, I discovered a small figure with flaming hair, bright red sweater and Land Girl breeches. Katy, at work. She was hitching an unwieldy log to the swingle chain while the chestnut stood quietly by, shifting her weight from foot to foot, ears and magnificent bushy tail flicking lazily. Stewart and the grey arrived. I saw him call to his team mate. A joke? An enquiry? I saw the smile on her face and a careless wave of her arm in reply.

The log secured, the girl checked carefully that all was well, then gave Mollie a brisk slap. A momentary check, as the pony took the strain on her capable shoulders, then the pair of them were merrily on their way over the hillside. Down they came, in perfect accord, girl leading pony and the log bouncing along behind. Mollie stumbled. The horsewoman held her firm, gave her an encouraging pat on the neck then continued leading her on down the rough track. The ex-Land Girl did, indeed, know her stuff. Peter could be congratulated.

Once on the road, Katy waved a cheerful greeting to me then got on with the job. Mollie was halted close to the men working the circular saw.

'This one's for you,' she called. And they gave her a hand unloading.

'You're doing well, Katy,' I congratulated her when she walked smiling towards me. 'Are you sure you can manage? It's heavy work.'

What do you think?' she retorted, quite indignantly. 'I'm fine. But you've been spying, Don! I saw your binoculars.'

'I'm sorry,' I replied, abashed by this honest reproach. With all the men up there, I wasn't sure whether you'd feel embarrassed if I came up to watch.'

'Why ever should I?' She looked astonished.

'I'll give Jimmie a good report,' I said, yanking my bike out of the ditch.

She grinned. 'You'd better.'

In a moment, with a cheerful wave, Katy and Molly were jauntily climbing the hillside and I was on my way to have a chat with the Boss. He expressed himself well-pleased with the experiment.

Good relations with Jimmie Reid, however, received a setback in March and for a short time I thought my job on the line. Towards the middle of the month we had a terrific gale. It was totally unexpected. The wind had been variable for a few days, seemingly working towards the dreaded spell of dry easterlies which would bring fear of fire. I went to bed quite early after a fairly tough day which had involved cycling from one remote part of the forest to another. Around midnight I was wakened by a blast of hurricane proportions which shook the house. It was weird, for a complete silence followed and I wondered if I had dreamed it. Nothing happened. I turned over and nearly fell asleep again. Five minutes later, the wind came howling from over the loch and hail rattled a merry tune on windowpane and roof. Sheet lightening lit up the window and the dark cloud high in the heavens behind. Seconds later, thunder rumbled in the distance. It was a cracker of a storm. The wind whistled through the ill-fitting door to my room. Roof slates lifted, slamming viciously back on the sarking and I heard at least two crashing on the cobble stones in the yard. Flickers of lightening showed up melting crystals running down the pane, and also a great crack in the glass. Hail and wind made so much noise they almost obliterated the sound of the sea but, between gusts, I heard it thundering on the shore and the surf surging back to meet the next great curling wave.

The sea! The launch! No warning of wild weather. No special precautions taken. Would her anchor hold? I leapt out of bed, dragging on a sweater, and ran down to the kitchen. Oilskins and boots were quickly put on and then I was trying to push open the door. Once through, it slammed behind me and I staggered across the yard to look over the gate hoping, against faint hope, that she was still riding the waves. Pitch dark, the only thing I could see were great curving walls of water rolling in to batter the shore – unimaginable, unrelenting force crowned with boiling white foam. Beyond, a restless loch heaved like the simmerings in a giant witch's cauldron.

She was not there! I was almost sure of it. I staggered round the little bay hanging on to whatever was handy. No luck. The *Mary Ann* was gone, broken adrift, vanished. Miraculously the dinghy had been carried over the shore and on to the ground above it. She lay on her side, apparently undamaged, her long line still holding her fast to an old fence post. I loaded her down with boulders, then retreated into the shelter of the house. Not to sleep. I hardly heard the final dying of the storm so busily was I thinking of the coming interview with Jimmie. Goodbye to a good report. Possibly goodbye Ardgartan.

It was a calm and changed world in the morning. I skipped breakfast and did a quick tour round the outside of my house. Broken branches were everywhere. One of the guardian pines had fallen, wrenched from the ground, its roots still embedded in soil. Seaweed, knee-deep, wet and slippery, plastered the loch-facing walls. One window had been broken by flying pebbles and the slates that had fallen lay shattered in the yard. But not as much damage as expected. The old fortress home had stood up well to its battering. In a now peaceful and idyllic bay, however, there was no sign of the launch. Where was she likely to be? With that wind, and on a rising tide, she would surely be swept towards the head of the loch. There were innumerable rocks upon which she could have foundered. Or, she could still be afloat, drifting inexorably with the now ebbing tide towards the open waters of the Atlantic. Almost worst of all, she might have travelled so fast in the night, she was now resting on the broad expanse of seaweed below Arrochar village. I would never hear the end of it from its good people.

My bike was safe in its shed. The next hour was energetic. Each small bay between Coillessan and Arrochar had to be checked and whenever I drew parallel with one, the bike was abandoned at the roadside and I hared for the shore. There, binoculars at the ready, I searched frantically hoping to find the missing boat but dreading, at the same time, to discover a calamitous situation. Each one was explored and found wanting. I became more and more puzzled. With that wind and tide, she had to have come this way! I reached the end of little bays and rocks and ahead was a relatively unindented shore all the way to the top of the loch. Not a sign of her. Now I was forced to make a detour which would take me close to the Forest Office and possibly my superior. Here, the Cally Burn meandered on its way to the loch. On an impulse, I propped the bike against a wall and took a look through a clearing in the trees.

And, there she was, the *Mary Ann*, high and dry on a bank at its mouth! She could hardly have chosen a worse spot!

I took a hurried look around. No sign of the forester. Good, I could discover the damage before reporting. Full of misgiving, I hurried along a well worn track down to the shore and five minutes later discovered that explanations would be required at once. The launch lay on her side, seaweed-bedecked and bows pointing out towards the loch. I noted her anchor missing. On the far side, hidden until this moment, Jimmie Reid stood leaning against the stern, arms akimbo and wig perfectly in place under his deerstalker hat. He was gazing fixedly at the spot from which I must appear.

'Good morning, MacCaskill,' he said, grimly. 'What is the meaning of this?' And he pointed to the sad sight beside him.

'I'm afraid she slipped her mooring in the storm,' I said, apologetically.

'Why was she not properly secured?'

'There was no warning of bad weather. No extra precautions were taken.'

Though it turned out that the boat was miraculously undamaged, except for scratched paint and a few dents in her planking, the next ten minutes were painful. Jimmie told me he had spotted the launch from his house and had already phoned for help. A fisherman in the village would pull her off on the next tide. But so far as he was concerned there could be no excuse for what had happened and I was lucky she had come to no harm. There followed a list of all the dreadful damage that could have been done, together with the probable cost of repair, then a few ominous hints as to what might happen to me as a result of my misdoing.

'See me in the Office this afternoon,' he concluded, at last.

I walked back to the road, retrieved my bike, and cycled to Coillessan to tidy up and consider my doubtful future at Ardgartan. Suddenly, of course, I knew most certainly that I did not want to leave the place. Surely to goodness I was not greatly at fault? Nobody had phoned a warning of bad weather and I was not a prophet. When I reached home, I found the squads, together with horsemen and ponies, already away to the forest. A bonus. The local bush telegraph was efficient and already they would be aware of what had happened. Belated breakfast, however, was not enjoyed.

At the end of the day, I reported to Jimmie Reid quite sure I was about to be given the boot.

'Are you feeling better?' my boss asked, a surprising twinkle in his eyes.

Good grief. What was coming?

'Fine, thank you.' I managed.

'You'll be glad to hear there's no serious damage to the launch,' he added. 'I'd like you to take her down to Mark to see Old John. Tomorrow. Take fencing material. Alec Seaton will go with you to check for deer damage.'

'Thank you, Jimmie,' were the only words I could find.

'Okay.' He actually smiled. 'I'll see you when you come back.'

All of a good night's sleep later, Alec and I were sitting in the launch chugging steadily along at a sober four knots.

'She seems all right,' he commented with a wicked smile. 'No serious damage done.'

'How did you know?' I asked, grinning.

'Oh,' he replied, laughing outright, 'nothing's sacred in this place.'

This was a beautiful morning, bright sunshine and a brisk, easterly breeze, wavelets chasing each other over dark waters, slapping restlessly at rugged shores, slapping happily against our bows. Colours were fresh and clean in the sunshine. A flush of green on the woodland trees which lined the loch suggested spring had already arrived – birch well-advanced, oak still to come. Behind, mature pine with warm, pink stems was topped with darkest green. Vegetation on the hillside still appeared bleached, though probably, the grass had begun to grow. Let everything be green as soon as possible, then we could stop worrying about fire!

The launch smacked her nose into the little white horses, her engine thudding out a regular paraffin beat, blue smoke blending with its acrid smell as it curled away over the stern. In her well were two rolls of netting, some wire and some stobs. Alec had his rifle on the seat beside him. We sailed past the heronry, now busy with the current nesting, and thought we must soon take a look. We cruised close to six Common seals hauled out on a small island. They observed our coming and our going apparently without interest. The tidemarks on their sleek and shining coats suggested total unconcern for the steadily ebbing waters which would soon leave them high and dry.

A derelict croft came in sight. In a small field enclosed by a broken-down dyke, the ridged, parallel lines of lazy-bed tillage could be seen. The cottage stood stark against larch and pine growing on the hillside behind. In front, a boulder-strewn beach sloped down to the loch and the flotsam and jetsam of many years lay rotting there. Salt-scrubbed boxes, staves, boots, bottles and rusting cans seemed a fitting stage upon which to set ancient timbers and crumbling walls. The whole was a reproachful memorial to another age when a hard-working crofter won a meagre living from this land.

'We're meeting your friends again,' remarked Alec, all of a sudden.

'Which ones?' I asked solemnly. 'I have so many!'

'Black ones.'

Right enough, at the far end of the sad little bay, hidden partially by an outcrop, stood a small group of the wild cattle I had met before. They were eating seaweed. Heads were lowered to crop the rich harvest and then lifted with untidy, straggling mouthfuls to chew. Wise eyes calmly regarded the rippling loch and they seemed unalarmed by our presence. But, there was another one I had not seen before. Startling red against the others but, until now, hidden behind a large boulder, an enormous bull stood contemplating the shore. A most impressive fellow with a proud head and wide, polished horns, he seemed quite unaware of our presence.

'He wasn't there before,' I exclaimed.

'New to me, too,' admitted Alec. 'Wonder where he came from.'

The animals all looked in good condition with thick shining coats and fat, well-covered frames. They had wintered well. It seemed there was only one bull, the red one. I turned for the shore and this was too much. As the launch came closer, the cattle all stopped feeding. Something triggered alarm. Led by the great red bull, they took off over the slippery weed across the shore. No stumbles. No falls. Surprisingly fleet of foot. Almost at once, they vanished into the forest and we did not see them again.

'Did you know,' asked Alec, suddenly, as the bay at Mark came in sight, 'that John always rows across the loch when he goes to collect his pay?'

'He told me,' I replied. 'I thought he might be pulling my leg.'

'Not a bit, and he won't have it any other way. The weather would have to be awful to stop him.'

And, it was not just across the water that the old man had to travel.

On the other side, he would tie the boat securely to an old bollard on the pier, dump his oilskins under a boulder, then walk about three miles up a track to the railway halt. There he would pick up a train that took him four miles further up the glen to Tarbet and when he left the train, he still had four miles to go to the office. For this last part of the trip he quite often got a lift from a passing motorist, but supposing he did not, by the time he had collected his pay and his orders, stopped for a gossip with anyone who happened to be around, and returned, he would have walked some fourteen miles to his boat and spent most of the day doing it. The return journey was often both hilarious and perilous, for various hostelries on the way each boasted old friends all anxious to have a chat and offer him a dram. On the final hop across the loch, Old John would be heard lifting his voice to the heavens in joyful song, his little boat pursuing an erratic course over the waters.

As Alec finished his tale, we rounded a point and entered the bay at Mark. The shingle beach, only recently uncovered by the tide, was distorted mirage-like in brilliant light and glared whitely at us as we sailed in. A broad ribbon of old seaweed, dead and black, etched an outline of the bay, contrasting darkly with this glittering radiance. The scene was idyllic but the cottage, in spite of the plume of blue smoke rising from its single chimney, looked somehow desolate. I felt guilty that it had been so long since my first visit, and anxiously scanned the house, its outbuildings and the ground nearby. There was no sign of John, or his dogs.

Searching the shore for signs of life, I suddenly realised that Alec was laughing. Speechless, he lowered his binoculars and pointed towards the far end of the bay. There, as on my first journey to Mark, was yet another of the black cattle beasts. This one, too, was jammed between two rocks but was not kicking frantically like the other had been. In fact, it was not doing anything at all. A swarm of flies rose and fell above the body in a frenzied dance of death, the cow's death. Alec was off his head. What was so funny? Then I saw it. A white and furry tail at the anus of the creature was waving to and fro in great excitement. A dog wagging its tail? Impossible. But, not impossible! One of John's collies had disappeared right inside the carcass by way of the anal passage.

Only its tail remained outside to give it away.

'Disgusting little brute,' I exclaimed.

'It's found the way to a damned good meal,' laughed Alec. 'I wonder where the others are.'

He clapped his hands, shouted, and had the answer at once.

The tail stopped wagging. Drooped a little. Then a squirming body, hindlegs, belly, forequarters, neck and head, all dripping blood and covered in dung, wriggled out backwards to see what was going on. We roared with laughter, the sound reverberated round the bay. At once, the gory head was raised. Eyes searched and found. A surprised yelp, and the dog took off across the shore. Now the climax. One, two, three, four more collies, all filthy, emerged from beside the corpse and raced away one after the other. The midday silence was torn to shreds by their yelping and we were killing ourselves.

We thought we would take a look at the unfortunate cow. Why had it died? I started the engine and gently brought the *Mary Ann* on to the shingle beach. As we searched for clues around the torn remains, for she had not been caught in the rocks and drowned, we noticed that the water rippling over the shore was discoloured and oily. Further examination revealed that all the surrounding seaweed had a film over it. The answer. A careless skipper had doubtless been emptying his tanks in the loch and, at this time of the year, seaweed would probably make up quite a large part of the cattle beasts' diet. Almost certainly, our cow had been poisoned.

Nearly low water, now. The launch would soon be high and dry. We decided to leave her where she was and, if necessary, wait for the incoming tide when it was time to leave. John must have been working in the plantation when he heard the dogs. Now, he appeared on the track leading to it. After giving him a wave, we picked up the rolls of netting, lifted them to our shoulders, then started off across the shore. His greeting, when finally we reached the path to his cottage could hardly have been called warm.

'You've come then,' he remarked politely, his eyes sliding towards the bundles we had dumped on the ground.

'The netting for your fences.' I stated the obvious.

'Hello, John,' Alec greeted the old reprobate. 'You're stuck with us a while.'

The old man glanced towards the launch, now resting comfortably on the shingle, and bowed to the inevitable.

'Come on in,' he invited, with not too obvious warmth, then led the way up the path.

It so happened that Alec had never before been inside John's cottage and, of course, neither had I. The old man was a widower and we were both curious to see how he managed for himself. The filthy collies greeted us exuberantly at the door and John made no effort to restrain their leapings and boundings. They rushed into the kitchen when he opened it and piled themselves in a steaming heap right in front of a smouldering peat fire. The stench was quite indescribable but, evidently, any state of dog was alright with John. He poked the fire into life.

'They're great company,' he explained, perhaps noting slight dismay. Then busied himself preparing a pot of tea.

We settled cautiously into an ancient settee which looked as though it was more often the sleeping place of the dogs. I looked around, curiously. Woodwork not painted these many years. Dingy wallpaper, once with a pattern, now an overall dirty brown. Patches of damp on the chimney wall. Smoke stain on the ceiling from a fire never out and, no doubt, a chimney only swept when the soot caught alight and there was a blaze. A window, never opened, boasted filigree spiders' webs spun from frame to lintel. The deal table, balanced uneasily on uneven flags, looked as though it had never been scrubbed. A pile of groceries in the middle – packets of cakes and biscuits, cheese and margarine, tins of stew, corned beef and spam, a box of matches – all had a kind of permanent air, as if once dumped there, they were never moved but simply nibbled at when needed. Products on the perimeter of this impressive pile would be gradually eaten away, but what treasures, or disasters, at the centre might be hidden beneath?

John seemed quite oblivious to any reaction we might have to his strange way of life and now we were here, delighted to have some company. Pipe in mouth and a trail of blue smoke lingering on the heavy air wherever he went, he pottered about happily, setting out mugs, warming a large earthenware teapot, extracting a packet of tea biscuits from his store on the table, and making enquiries about the various friends he sometimes met in the Forest Office.

I made an excuse to go and see whether the launch was safe, but really to get a lungful of fresh air. A door opposite the kitchen was ajar and I took a peep around it. Here was another surprise. This was the parlour. Spotless, not a thing out of place. Printed curtains neatly drawn to exclude most of the light. An old carpet dotted with sheepskins, perhaps to conceal worn or stained patches. A new rug in front of the fireplace. A stick fire laid, but probably never lit. A three-piece suite, shiny brown hide, surely never sat upon. An imposing standard lamp, with rose-pink shade and tasselled fringe, standing sentinel in a corner. Knick-knacks on the mantlepiece, all in neat order. A small bookcase with an interesting mix of leather-bound classics and a number of tattered Penguin paperbacks. On a small, occasional table, centre stage, the faded photograph of a young woman. All seemed to wait, in a state of suspended animation, for the home-maker who never was there.

Outside, breathing in clean air and revelling in the bright sunshine, I struggled to get into perspective this lonely old man and his solitary way of life. I had intended to ask Jimmie whether another cottage might be found for him nearer the village – if anything serious happened to him here, no one would know it, perhaps for many weeks. But, could he be persuaded to move? Now, as I thought of the spotless parlour, the homely kitchen, and John happily dawdling amongst his odorous collies with the tea things, I knew it would be a great mistake to uproot the old man.

In the living room, tea was now steaming from three mugs on the dirty table, and a neat plate of biscuits awaited our pleasure. The dogs, comfortably full of their disgusting meal, slumbered peacefully in front of the fire. Alec and John lighted up, then for ten minutes in the swirling smoke we sipped our tea and discussed seed potatoes and other crops the old man might grow.

We had come on duty, however.

'We'd better take a look at your fences, John,' I said, at last, knowing fine that all would not be as it should and I would have to give the old man a ticking-off.

'Aye,' he grunted and postponed the evil moment a little longer by clearing away the crocks.

We sauntered up to the first of his two little fields. They lay side by side behind the cottage, rough grazing ground sheltered by the forest. Originally, the old stone dyke had enclosed them but, beautifully constructed a hundred years ago, or more, it was now falling apart. John's fences, taken round the outside, were designed to cover the gaps and to keep the sheep in. It was no surprise to find no sheep in the first.

'Where are they, John?' I asked sternly, as we leant on the wall and gazed into emptiness.

'Maybe in the other field,' he suggested. 'It looks as if the gate's open.'

But Alec had already wandered over to look and now had returned. He shook his head solemnly. 'They're not there, John.'

'I must have missed a hole,' the old man apologised. 'I was blocking them with stones.'

The old rascal managed to look thoroughly upset over the sins of his sheep, but I guessed nothing at all had been done since my last visit. At this time of year there was little grazing available but I could see no signs of supplementary feeding. Survival was in the forest.

'It's not just the fences,' he suddenly blurted. 'There's a problem.'

'What do you mean?' I asked, wondering what the next excuse might be.

'Wee beasties!' replied John mysteriously. 'They're killing the grass in the other field.'

Though it was still early in the season, our mild climate often meant that the grass was beginning to grow by the end of March. When we arrived to look over the gate, we found large patches where it was thin and turning brown. In addition, small fluffy balls of old vegetation were drifting on the air like puffs of dandelion seed.

'It happened last year, too,' John said. 'Grass was coming in nicely, then it was turning brown for no reason I could think of. It all died. Must have been the beasties.'

I could not help smiling. John obviously thought that leprechauns, or other mischievous little people, were casting a spell over his croft. In a way he was right. We began to examine the ground. Everywhere, patches of grass were dead or dying, eventually leaving the soil quite bare. This was devastation to a man farming sheep. There had to be a reason. With my foot, I idly scraped away the soil, looking for clues. Then, suddenly on a time warp, I was back in the Forest of Dean at a lecture given by my old mentor, Professor Crystal. That was it!

'Come over here,' I shouted to the others and knelt for a closer look. There they were, in their hundreds, the 'wee beasties' John was afraid of. Brown heads, curled wrinkled bodies, three pairs of legs and two rows of parallel spines on the ventral surface: cockchafer larvae, or *Phyllo pertha horticola*. The roots of grass were a favourite food. Of course, it would be dying.

'You were right about the 'beasties', John,' Alec laughed as they joined me.

'Aye,' he replied, ruefully examining the scraping I had made. 'So, why are there all these bits of grass blowing about?'

'I'm not sure,' I said, wishing I had a magnifying glass to see the tiny creatures better.

We each set off over the field to have a look at the work of the industrious insect.

'Come up here, Don,' called Alec, who was pointing triumphantly to something on the ground.

Three dark, purple droppings lay neatly in a pile on one of the bare patches.

'Badger!' I exclaimed. 'Well done.'

Badgers are fond of juicy mouthfuls of any beetle. The larvae, too, are enjoyed. A family was obviously paying regular visits to John's field. There must be a sett at Mark. The little puff balls, floating about on the breeze, were probably caused by busy badger snouts. They would nose into the soft, sandy topsoil, pushing aside and turning over the dead grass until it was woven into tiny balls light as thistledown.

'Well, I'm damned!' exclaimed the old man. Then, anxiously. 'Will Alec shoot them for me?'

'Heavens, no,' I remonstrated. 'You ought to be pleased to have badgers, John. They'll help to get rid of the beetles.'

Time was marching on and we each still had another job to do. Alec
was to investigate deer damage around Mark and I wanted to make a
quick dash to a cliff further down the shore. It rose sheer from the sea to
a couple of hundred feet and there might be a raven's nest on one of its
many ledges. By now, if there were young, they would be hatched.

'About the sheep, John,' I said, as we turned to leave. 'I am sure
Jimmie will not allow of any more excuses for letting them into the
forest. You've got far too many, anyway. Why don't you sell some?'

The old man looked glum. 'I couldn't do that,' he said. 'Sheeps is
cheerier than trees.'

We thanked him for our tea, said we would be down again soon, then
left him to contemplate the latest of his problems.

'I'll see you in an hour or so,' I said to Alec, and we went our separate
ways.

It was a beautiful nest, its sticks and twigs all cunningly woven
together in the safe far corner of a broad ledge, protected both from the
direct heat of the sun and rain from the prevailing westerlies. I had
nipped up through young forest to the top of the escarpment and had
worked along it until I thought I was above the likely spot. Indignant
'cronks' from a pair of ravens coasting anxiously overhead suggested I
was. A steep slope of grass would take me down and a little closer. I lay
on my belly and wriggled. At the point of no return I craned my neck,
as far as I could, to look over. And there they were, in their snug home,
three black ugly 'ducklings', a long way from the beauties they would
eventually become. Enormous bills were agape revealing red throats
waiting to process the next meal their parents would bring. But, what was
this? On a small ledge beneath, and to the right of the nest, another
chick. Fallen out? Too small, as yet, for that. Shoved out by more
dominant siblings? Possible, I supposed. More likely, it had been
accidentally shunted out by one of the parents and was extremely lucky
to have been safely caught.

What to do? The youngster looked thin and listless and was lying,
rather than standing, on its insecure perch. I didn't think the adults were
feeding it and it would perish if not rescued. How to get there? A sapling
rowan grew from a crack in the rock face, midway between me and the

bird. Just possible, I thought, and squirmed down to it feet first, tried it out, reckoned it was safe. I grabbed a branch with one hand and with the other just managed to lift the waif from its ledge. There was no way I could put it back in the nest so I popped it into my rucksack. The return journey was hair-raising but, at last, it was done. I took a quick look at the youngster. Its eyes were closed, but it was breathing. Well. It was worth a try.

Twenty minutes later, I was back beside the *Mary Ann*, now floating in eighteen inches of water. In a few minutes, Alec came striding out of the wood. Quickly removing our boots we waded the few yards to the boat and climbed in.

'Are you poaching an egg?' he asked, noting how carefully I laid my rucksack in the stern.

'Not quite!' I laughed, and picked out the small bundle. 'This is 'Cronk'.'

Back at Coillessan, the launch safely secured against anything the weather could do, Alec organised tea while I made a temporary home for my raven in an old tea chest. I lined the bottom with plenty of newspaper, arranged a perch across one corner and another behind the netting door in the front, then placed the bird on plenty of wood shavings to keep it warm. I found some raw mince in the larder and, at first, had to prize open a reluctant bill and poke small pieces down his throat. It soon got the hang of it and was gaping greedily for more.

'Thanks,' I said to Alec, as he handed me a mug. 'By the way, I forgot to ask. What's the deer situation down there?' 'Too many,' he replied. 'I'll need to spend a few days.'

'Keep an eye on that nest for me.'

In due course, he was able to report that three young ravens were fledged and flying over the cliff. The adults were there, too, and the family were performing spectacular aerobatics, swooping and soaring over sea and hillside, and 'croaking' raven messages to each other all the while.

Cronk was to stay with me for nearly a year. I decided to assume he was male, though I never knew for sure. He was a survivor and had no intention of dying. Any food I gave him was swallowed greedily and he grew, day by day, almost visibly. In a week he was perching on the sticks in his cage and I took to leaving its door ajar. The moment came. Suddenly he took the plunge and flew with a hurried, desperate flapping of his wings to the top of the open kitchen door. My chick had fledged. He looked rather surprised, then began preening the feathers on his breast as if flying was something he did all the time. From then on, whenever I entered the room, he would glide straight to my shoulder and perch there, companionably poking my ears with his bill or muttering 'pruk-pruk' conversation as I went about my business.

It was inevitable that the bond between us grew strong – I was his parent. He began to accompany me on trips into the forest, at first perched on my shoulder, but bolder and bolder, taking off for soaring flights to the tree tops. I often thought him lost, but whenever I set off for home, a handsome black bird, his feathers glossy with good feeding, would fly to join me again. Once home, he would settle quite happily on a favourite perch, usually on top of the old kitchen cupboard, until food was offered.

Cronk was a great character and as he grew to adulthood was always getting me into trouble. He had complete freedom and eventually was only coming home in the evening, announcing his arrival with a peremptory tapping on the kitchen window. His favourite ploy was to remove pegs from all the nearest washinglines – what was a few miles to a raven? Irate housewives were liable to find their beautifully clean laundry on the ground. He was not above stealing ice cream cones from the village children. Just as the longed-for delicacy was being lifted towards an eager mouth, it was snatched away and dropped to the ground, uneatable now, only a few yards away. Many a time I had to put my hand in my pocket to console a wailing infant. The ultimate outrage was to pinch a five pound note straight from the outstretched hand of a lady paying the man at the grocery van.

The end of our association was inevitable, of course. First of all, Cronk began to stay away overnight and I had no idea where he was roosting. For a little while, from time to time, he would come flying to my shoulder, when I was in the forest, to greet me with the usual companionable noises. But the intervals grew longer and, one day, l realised he had not made contact for over a week. That was it. I never saw him again. Perhaps he had found a mate.

six: fire in the forest

IN April the weather became warm and dry and Jimmie caught spring
fever! Normally, I saw little of him but now was summoned to his office
more frequently. He looked ill, was gaunt and heavy-eyed, dark circles
beneath his eyes. The disease was called 'fear-of-fire' and most foresters
fall victim to it in the spring, passing sleepless nights imagining their forests
blackened and burnt and Authority wanting to know why. Each one was
affected according to his nature, at best becoming mildly irritable, at worst
a nervous wreck. Our boss belonged to the second category. 'Have you
checked all the fire traces?' he would growl. 'Are fire-brooms in position?
Is the roster drawn-up for the weekend?' It all seemed a bit ridiculous to
me but then I had never experienced a fire. That was to come.

It was now the planting season, a particularly busy time in the forest.
A new plantation was taking shape out on the hill beyond Coillessan and
for the past few weeks, whenever we were free of frost, Archie had been
at work on his tractor, ploughing over the peaty contours – ground heavy
with moisture and the job a long one requiring skill and endless patience.
A squad would soon be going up there to do the planting. Ardgartan also
boasted a tree nursery and the spell of good weather had brought perfect
conditions for working the soil. I sent Bill O'Connor there with a large
squad of men and Katy was prised away from her beloved Mollie to
supervise a team of women. Men were also still working in the Douglas

Five hungry mouths to feed but with asynchronous hatching, the smallest hen harrier chicks will probably be eaten by their larger siblings if there is a shortage of prey.

An unusual visitor to the nest, the male hen harrier did not bring prey.

The kestrel cock brings prey to four healthy youngsters on the cliff eyrie.

An ingenious hoodie nest on an island bare of trees.

An angry call from a raven.

Feed us! Hungry raven chicks with mouths agape.

The magnificent capercaillie cock defends his lekking place.

His mate nests on the ground and is beautifully marked to merge with the forest floor.

On a cliff ledge high above the sea, the fulmar hen had laid her single egg.
She ejected a foul-smelling 'missile' from her mouth at the photographer!

Guillemot and razorbill on a crowded cliff. Is there room to let?

A formidable predator on cliff and shore life, the great black-backed
gull is also a useful 'cleaner-up' of carrion.

In a gannet colony, where every available space on a cliff is occupied,
the parents always fly back to their own nest and chick.

A well-camouflaged, ground nesting bird, the curlew sits
tight on her eggs and will fly from it only if disturbed.

The golden plover also merges with the surrounding vegetation
and is difficult to discover on her nest.

This heron has caught fish in a nearby loch and has flown in to feed her young.
Her mate will have been on guard while she was away.

The little dipper worked hard for this beakful of insects – its family was
in a cunningly constructed, domelike nest beneath a nearby bridge.

A large fish for osprey chicks.

The red-throated diver nests close to the water
and is often disturbed by fishermen.

Puffins can only be photographed in the breeding season.
They spend the rest of the year at sea.

Old Scots pine – a survivor from long ago.

A walk in the woods beside Loch Tummel.

A lonely pine on the way to Ben More.

Autumn birch in Glen Lyon.

Dawn on Loch Lubnaig.

Winter among the larches.

A remnant of the Old Wood of Caledon.

Bitter winter on Rannoch Moor.

A cluster of sitka spruce cones.

Evening on Loch Teacuis.

Wood where the thinning programme was coming to an end.

The ploughman was needed down in the nursery as soon as possible, so one afternoon I cycled up to the plantation to check on progress. It was another fine, dry day with the wind in the east and I chuckled at the thought of Jimmie pacing up and down in his office as he worried about impending disasters. But new growth was showing on the banks of the road – primroses and violets nestling in new grass, heather foliage coming to life – and above the forest edge a faint flush of green was edging out the bleaching of winter. Soon the anxious forester could stop worrying for there would be little dry vegetation for any fire to feed upon.

I dumped the bike and began the walk up the hill towards the ploughing. Archie's black sou'wester, worn whatever the weather, was bobbing up and down, swaying this way and that, as he jolted across the rough ground. I noted, with satisfaction, the long beautifully cut furrows, the black peat turned to the sky. He saw me, waved an acknowledgement and, in a few minutes, his monster machine came roaring towards me along the next furrow. The ploughman switched off the engine and I watched his burly figure, fresh complexion brick red in the biting wind, hop down to stretch his legs.

'Afternoon, Archie. You're nearly finished, I see.'

'Aye, Don. It'll be done tonight. I'm glad you're here. There's something I want to show you.'

'Oh?' I enquired, but his blue eyes twinkled and he made no reply.

We stumbled across the furrows towards a small rush-covered area left unploughed right in the middle of the huge expanse. I had assumed the ground unsuitable for some reason or that he might have discovered one of the ancient, historical sites with which the area was littered and which I had ordered should be left undisturbed.

'What is it?' I asked, again.

'It's special,' was all he would say as we drew nearer.

Right at the edge of the rough patch, which looked perfectly ploughable to me, he put his finger to his lips and we positively tip-toed towards its centre. Then he halted.

'See that gap in the rushes?' he whispered, pointing further ahead. 'Take a look.'

It was only a few feet to go and I crept forward with care, wondering, half-guessing what I was going to see. And there she was, in a hollow on the ground lined with bits and pieces of grass and heather, a curlew sitting tight on her eggs. It could have been either sex, for both share incubation duties, but I settled for it being the hen. Brown speckled feathering blended well with the dead vegetation and between parted feathers on her breast I could just see two of her eggs, dark brown spotting on a paler brown, and knew there were probably two more tucked away behind. The downward curve of her bill made her look both lugubrious and severe, and beady eyes regarded me with dawning alarm. I retreated as quietly as possible and she did not fly. Perhaps she was used to Archie.

'A nice domestic scene!' I agreed, smiling, when I had rejoined him.

'She's a great wee bird,' he commented, enthusiastically. 'I've been driving up and down quite close for a fortnight and she's never stirred. Was it all right to leave her alone?'

'Of course. I'll tell the men to keep clear when they're planting. She might just stay the course.'

We walked back to the tractor and, after a few words of con-gratulation on a job well done, I told him to report to Bill O'Connor the next day.

Cycling back along the road to Coillessan, I met Alec in his van returning from a stalk above Mark. I had been going to get in touch and now seized the opportunity.

'Had any luck?' I greeted him.

'Aye. I got a buck. A nice six pointer. Saw some stags, too. Some good yins coming.'

'I'll look out for them.'

The roebuck grows his new antlers during the winter, but the red deer stag casts his old ones in March/April and would only just have started growing in the new. Alec will have been on the look out for promising 'heads' and likely 'Royals'. Though his job was to control the numbers of deer in the forest his instinct, as a good stalker, would be to weed out inferior beasts first and to only kill really good ones if there were too many.

'You and I were going to have a day on the hill,' I reminded him.

'I hadn't forgotten. In fact, it's nearly time to look at the dens. We could take a walk up to the one in Glen Croe tomorrow evening, if you like. I need to check whether it's being used.'

Suddenly, I remembered what went on during April and May and was not at all sure I wanted to go. I hesitated. Alec was taking the first step towards that traditional annual ritual, a visit to a fox den that would mean, in the end, the slaughter of beautiful and intelligent wild creatures. This slaughter would be repeated at the number of dens he could find occupied in the forest area. It all seemed completely unnecessary to me and I wanted no more part of it than was relevant to my job, namely to know that it had been done. However, the stalker would be leading me to a den I had not so far discovered. If it was one in use, maybe I could watch it, at least for a few weeks. Perhaps we could talk out the problem, too.

'Okay,' I agreed. 'Meet you at the end of the road about six.'

'Right. I saw you talking to Archie,' Alec continued, as he prepared to carry on.

'He was showing me a curlew nest,' I told him.

'I guessed,' he chuckled. 'Did you know he was showing one to Jimmie Reid at the same time?'

'What!'

'Use your binoculars. Pretend to be watching a bird in the sky, or something. Take a look the other side of the loch, on the main road about three hundred yards from the village.'

He was right. Jimmie's battered old Morris was parked on a bend. The driver's window was half open and his red wig, askew, could just be seen above a telescope balanced on the rim. He was crouching to the eye-piece and was clearly focussed on our part of the hillside. In an uncomfortable moment, I reckoned we were focussed on each other!

'He's getting jumpy,' explained Alec. 'And we're close to the weekend.'

I knew what he meant. It was actually only Thursday but our boss would already be in a state about the walkers, the men and women who came to climb the nearby Cobbler and camped in the caves above the forest. They all lighted little fires to cook their evening meal and he thought one must surely burn out of control. Many of these enthusiasts went to the pub in Arrochar in the evening. Jimmie would be there, ahead of them, hoping to learn their intentions in regard to camping and to discourage the lighting of fires. If this failed and, later, camp fires twinkled merrily on the hillside, he would send someone up to see them put out.

'He can't get about as well as he did, so uses the 'scope,' continued Alec. 'He's crazy about his trees and if there's a serious fire, he'd never get over it.'

'I see.' I said, absorbing yet one more facet of Jimmie's strange character.

The next morning I went straight to the nursery. This was ideally situated on a long, gentle slope to the loch, was well-drained and also collected any sunshine that our wet, west coast climate might allow. It was completely surrounded by mature forest and so was sheltered from the worst of damaging winds. We grew, from seed, all the trees that would become the forest of the future – Scots pine, Sitka and Norway spruce, Noblis and Douglas Fir, Japanese Larch, and a few others. Tiny seeds would take root to become seedlings, seedlings grow to be saplings, and the saplings, according to the terrain in which they were to be planted, would be ready to go to the hillside at two, three or four years old.

The men and girls all had their allotted tasks in the nursery and they would be working there for five or six weeks. It was a beautiful day, a drying east wind, the sun already quite hot for the time of year. This could cause problems. It was vital to keep the roots of the small plants moist while they were being transferred from seed beds to transplant lines, and again, out to the hill. If they dried out, they would die.

I went to the seed beds first where Old Dougal, expert axeman, was in charge. Archie had previously prepared the ground – three foot wide parallel lines of raised soil, each divided from the next by a narrow walkway. I found Dougal, with two other men, raking the soil into a fine tilth ready for the seed. He was meticulous in seeing that it was properly broken down and I knew he would put as much of his expertise into planting embryo trees as ever he did into the growing of his beloved onions. The usually silent old man enjoyed this work and as I approached I could hear him gossiping and joking with the others.

'Morning, Dougal,' I called. 'How's it going?'

'Fine,' he replied. 'Need to get on, though.' And, with a quick glance at the hot sun, he hurried to continue with the job.

I watched them for a few minutes. They were nearly finished on that strip and I knew that, in a short while, they would be scattering the seed over it. After that, as quickly as possible, they would carefully rake fine grit over all so that when the rains came, the seed would be free of clogging soil. This was the start of a growing process that, in the right conditions, meant one pound of seed would become one hundred thousand seedlings. It was important that they got off to a good start.

Nearby, at a bed of year-old Sitka spruce seedlings which, at a glance, all looked good and healthy, two more men were digging up the tiny trees, collecting them into bundles of roughly a hundred, then putting them into pails for collection. These were to go to the women's team to be placed on the transplant boards. And this was my next stop.

Behind a large hessian screen, a somewhat rickety affair which protected the plants from wind and sun and could be moved about as necessary, another scene of hectic activity was taking place. Katy and her squad of women were transferring the one-year-old seedlings brought from Dougal's seed beds, on to transplant boards. Each, when full, would be taken to another part of the nursery for planting out – space and light essential to promote strong growth. The whole operation never ceased to fill me with admiration for it required precision, speed, perfect timing, and absolute economy of movement.

This team, Katy in charge, was expert. As I arrived two 'nippers', lads who were allowed to do unskilled jobs in the forest, had just deposited pails of seedlings beside the transplant boards and the girls were swinging into action.

'Morning, everyone,' I said, but not a head was raised. And, again, 'How's it going?'

'Okay, Don,' replied Katy, cheerfully. 'But we can't stop!'

The transplant boards were roughly eight feet in length with slots in them at two-inch intervals. When one was filled, another matching one was folded over the top to prevent damage to the tender plants. Metal clasps were then snapped over the hinges to make sure they stayed closed. Each girl picked a handful of plants from one of the pails. She held the bundle in one hand and with the other pushed a single seedling into a slot on the board so that its roots fell free beneath. As she finished each handful, another was seized. So fast did all their nimble fingers work, hands flying over the boards at an amazing speed, teasing and gossiping all the while, each would have handled some five thousand seedlings by the end of the day. I noticed one of the women ruefully examining her fingers.

'Problems, Morag?'

'It's just those bloody Sitkas!' she explained, and showed me scratches, torn and bleeding, from the stiff, prickly little seedlings.

'Do you want to get them attended to?'

'You must be joking,' she replied, and returned to the board.

Feeling somewhat superfluous, I left them all to it and began the walk over to the transplant lines. Here were three large sections where Norway spruce, Scots pine and Japanese larch were doing well. A fourth area had been prepared to receive the one year-old seedlings from the girls, and it was to this that I made my way. Bill O'Connor was in charge.

'How's it going, Bill?'

'Fair, only fair, Don,' replied the lugubrious Irishman. This was standard. He was always gloomy in the morning, even when all was going well and the weather was fine. Katy said it meant the latest applicant for the post of wife had declined the honour and returned to the city.

Ten men were at work here, and it was another precision operation. Two men were running from the transplant area to the hessian shelter in order to pick up another loaded board. Two were walking as fast as possible from the shelter to the transplanting area carrying another. Two were bent double placing a board in position on a long, shallow furrow. Archie was sitting on his tractor at the end of one of the lines waiting to go into action.

I watched the pair with the loaded board. They laid it carefully, end to end with the next one, roots hanging free, and secured it in position with metal pins knocked into the ground. This happened to be the last one in the line. Now Archie, on the small tractor, started along it, driving slowly and carefully, turning in the soil over the fragile roots and the men followed behind, tramping it down as firm and close to the board as possible. When they reached the end and all were secure, the soil was levelled off to the width of a spade and the retaining pins pulled out, metal clasps released, and the boards lifted away. Hey presto. An immaculate line of tiny trees, all firmly planted. Nine inches away, the next furrow would be marked with a length of twine and a line cut with a spade to the depth of the roots of a new lot of seedlings.

Nearby, two more men were forking up three-year-old Norway spruces, tying them into bundles of a hundred and loading them carefully into the forest lorry which was standing by. Once full, a tarpaulin would be placed over all to prevent drying out and the driver would be off to the hillside. It was all backbreaking work and Bill O'Connor had his team working flat out. Occasionally, a man would pass a word to another, but mostly they worked in silence, knowing the drill so well. Our Irishman showed his strength in keeping them cheerful in what was, after all, a pretty dull, repetitive occupation.

'They're going well,' I remarked, by way of congratulation.

'They'd better,' he retorted.

This did not sound like the usually good-natured ganger who though a strict disciplinarian was never unreasonable.

'What do you mean,' I asked, curiously.

'He's at it again,' he replied with a wink, and could only mean one thing.

I turned to look over the loch and there, sure enough, was the forester's old Morris parked in the same place again. Through binoculars I discovered the famous telescope, this time balanced on the roof, the man on the far side propped against the door.

'Better get on, then.' I smiled, but made no comment.

My bike and I hitched a lift on the lorry and, as we left the nursery, I looked back on a nice scene of industry in which all was running smoothly, and was reminded of a colony of ants, in slow motion, all

busily carrying out their allotted tasks. As we bumped over the rough
forest road there was plenty of time to think about the strange anomaly
that was my boss. Here was a chap who seldom bothered to look at his
forest except from various vantage points through a telescope, but who,
according to Alec, would never get over the shock of a serious fire. He
would mourn each blazing tree as if his own limbs were being devoured.
In addition to this, the evidence I saw about me suggested a
silviculturalist of some distinction who was genuinely interested, apart
from his responsibility to his superiors, in the welfare of his trees. The
newer areas of forest were well planned to contain a mixture of species,
each to grow where it was best suited, and he had inherited an excellent
mature one which, he saw to it, was watched over and managed with
care. It was, in fact, an excellent forest for a young forester to learn the
job in.

As we trundled to a halt on the road, I saw dark peat furrows
stretching over the hillside for ever and little men scattered all over them,
most bent double! Within the next six or seven weeks, some 300,000 little
trees, of mixed species, would be planted, each man probably achieving
about 1000 in a day. A huge operation. How could they ever finish? And
yet they would, the task undertaken in all weathers except a sudden and
unexpected frost. I gave the driver a hand, lifting out bundles while he
carefully laid each into a 'sheugh' beside the road and covered the roots
with soil.

A weird and wonderful sight, came slithering down the bank beside
us. It was the pony, Jock, with his horseman, Dick. This time there were
no dragging chains rattling along behind and instead of the usual
harness, he was rigged out in a curious sort of iron frame, padded to
prevent sores. This fitted over his broad back and extended on either side
to form a pocket, a containing space into which trees could be placed.
The whole contraption was held in place by straps secured beneath his
ample belly.

'How're they doing?' I asked Dick as they halted beside the lorry.

'As usual.' He laughed. 'This is old soberside. Tommie is dancing all
over the place and has nearly capsized a couple of times.'

I watched the horseman load up the placid Jock, taking bundles of trees from another sheugh, putting them into hessian bags and then the bags into the empty 'pocket' on either side of the pony. He finished off by piling as many as he could on top of the pony's back and securing them all with a rope. The order was given: 'git up there, boy', and then Dick was encouraging him up the steep bank. For a few minutes, I watched an unwieldy mountain of hessian, with a magnificent brush of a tail and four sturdy legs, plodding up the hillside, a man walking alongside. The ponies would match progress with the men, depositing their trees in the moist bottoms of the peat furrows at suitable places across the hillside. From there they would be collected for planting.

Peter came down with Tommie. Andy Murray, the ganger, was with them.

'Afternoon, Andy. How's it going?' I asked the inevitable question.

'Well,' he replied. 'But I wish it would rain.'

I was worried, too. We had now had a long spell without it and the soil would be drying out quickly on the well-drained slopes. If drought persisted, thousands of newly planted trees would be at risk and many hours of work, together with a great deal of money, would have been wasted if they died. Jimmie would be off his head! As we stood looking, I thought of each small tree being planted, teams of three men working along the furrows, the drill the same as that I had learned as a young forest worker. A vast hillside. Thirsty work in the warm sunshine, men sweating and cursing. It was a tough job and boring, too, unless you had a vision of the future as Old Dougal might have had. I thought, with him, of a fine forest taking shape which would clothe a barren hillside, would hold the precious soil in place and provide shelter and a suitable habitat for a wider range of birds and mammals. That was easy. This time, I wasn't doing the hard work!

'How's Tommie doing?' I turned to Peter.

'He'll settle down. One day!' he remarked of his enthusiastic work mate. He laughed and patted the offending animal. 'By the way,' he added. 'Tell Archie his curlew's still safe!'

It was good news for the ploughman and reward for the trouble he had taken. But for now, there was one more job to do. Jimmie had announced a rare visit to the Douglas Wood and I needed to be there.

He would be 'up' sometime that afternoon, he had said, to see the last of the thinning. Tiresome old bugger, I had thought, you know it's a copybook job. Why bother? Another thought occurred. Did he really think my men would be dropping lighted cigarettes all over the place on a tinder dry forest floor? Then I remembered it was Friday. Of course, campers! They could easily pass through the wood. The Boss would be prowling around for likely sinners on their way to the hill.

Andy had his bike with him, too. He was due to return to the thinning operation. We cycled back along the road to the nearest point to the Douglas Wood and then climbed the hillside. When we arrived, the men were in the middle of the lunch break, sitting with their backs against the tree trunks, chatting over mugs of tea and smoking. But, all at once, as we moved towards them, the talk began to die away and then there was silence. It was not on my account. They were looking in the opposite direction. A man came slowly walking through the wood, from the other end, pausing from time to time to gaze all about him at the trees. It was Jimmie Reid.

'Afternoon, boys,' he said, cheerfully enough and nodded to me.

'Aye, aye,' some muttered, as they rose hurriedly to their feet. Last drops were drained or emptied to the ground and each, grinning self-consciously, stamped out his offending 'smoke'.

'How's it going?' he continued, a rare twinkle in his eyes as he noted the action.

'Fine,' I replied. 'It's looking good.' And was proud of the job we had done.

But, suddenly, Jimmie was not listening. He was urgently sniffing the air.

'Smoke!' he exclaimed. 'Don't you smell it?'

Suddenly, I did. Deadly to a forester, it came from the east, a faint whiff only and elusive on the breeze. There was nothing to be seen, though. Now the whole gang was sniffing, too, and looking for tell-tale signs of smoke through the canopy. In the hope of seeing better, I began to run to more open ground, but Jimmie had a better idea. He had dashed to the nearest tree and, astonished, I stopped to watch. He 'squirrel-ed' up to the lowest branches, no mean feat for there were no foot or handholds, worked quickly through them, impatiently brushing aside needles and tearing at clothing that had caught. Then he began bouncing up and down on a branch in order to gain height for a leap and a grip on the next above. That achieved, he vanished from sight into the thick canopy and all we could hear was vociferous swearing. I noted his wig, bedraggled and forlorn, hanging from a twig.

'Does he always do that?' I asked the men.

'Oh yes,' they chorused. 'Standard practice if he thinks there's a fire.'

I thought they were pulling my leg.

'It's the Black Planting,' came a muffled voice from on high. And down he came again, slithering, sliding, crashing through the swaying branches. He collected his wig on the way, unashamedly pushed it into his pocket, then jumped the last few feet to the ground.

'Come on, boys. Get moving,' he yelled, and set off down the hillside.

I could not believe this. Jimmie bounded over that hillside like he was a mountain goat, leaping dead branches and ditches, dodging obstacles in his way like he was a winger running for the line, and, eventually, sliding on his seat down the final bank to the road. Getting on a bit, my foot, I thought as we tore after him. The old car had been parked out of sight round a bend. The forester positively sprinted along the road, threw himself into the driving seat, slammed his wig into place and anchored it with his deerstalker. We clambered aboard, some inside, some on to the running board, clinging to whatever was handy. Jimmie crashed into gear and kangarooed away. We roared along the rough road swerving to avoid potholes, sliding precariously round the bends, and missing a ditch at the side by the narrowest of margins. Dust rose in clouds behind us.

We reached the junction to the main road and at once saw where the fire must be burning. A huge pall of smoke hung over the Black Planting and dense dirty clouds were rising to feed it. Of the actual blaze there was no sign as yet for trees on either side of the road hid it. We spun triumphantly on to smooth tarmac and accelerated towards apparent disaster.

'Can you see the fire?' shouted Jimmie over the outrageous noise of his engine.

'No,' I yelled back. 'Round the next bend, I think.'

'Hang on.'

If it was possible for the old banger to go faster, it did, and as the alarming situation drew too quickly near, I'm sure each one of us was preparing for the worst, frantically going over in his mind the prescribed drill for an occasion such as this. Over the next rise and down again, round a never-ending bend, and there, at last, was the fire – and the cause of it, too. In the middle of the road, an ancient jalopy was well alight, black smoke swirling and curling around it, flames roaring as they fed on escaping fuel and consuming everything that would burn. In the middle of the inferno, a man was desperately trying to stamp it out.

Jimmie swore, crashed on the brakes and skidded to a halt. The situation was potentially serious. Dry vegetation on the edge of the plantation was well alight and, worse, the fire had already spread into the young trees. Quite a few were burning.

'Christ!' exclaimed Jimmie, taking in the situation at once. 'The wind!' It was blowing from car to forest and spelt disaster.

The owner, a grimy, smoke stained individual came running towards us. It was Bob Daniels, from the village.

'She suddenly blew up,' he explained, apologetically.

'Get busy, Don,' shouted Jimmie, ignoring him.

'Beat towards the rocks,' I shouted to the men, suddenly realising that a large outcrop could form an effective barrier between the fire and the major part of the forest. 'Lend a hand, Bob, and do as I say.'

Brooms were on a fire stand nearby. We worked in line ahead, handkerchiefs, scarves, anything over our noses and mouths, three teams of three men each. Each leader was first at the fire, beating hard, using all his strength to extinguish the greedy flames. When he could stand heat and smoke no longer, he retreated to the rear for a spell and everyone moved up one. Number Two attacked any remaining flames still licking at the smoking vegetation, then advanced to the next big blaze. Number Three came along behind, making sure that the fire did not break out again. And so on. Thus, in turn, each man fought the worst of the blaze with the maximum of his strength and energy. We acted fast, beating, stamping, cursing, gradually creating an arc of burnt out vegetation between the main blaze and the rocks – and, thus, a fire break. It was tough. Smoke stung our eyes till the tears streamed, and had us coughing and retching. It was unpleasantly hot and the men up front could only stand a few minutes at a time. However, the adrenaline flowed, anger spurred us on, and I do not think anyone had time to be afraid.

Gradually the hungry fire was angled back towards the outcrop. The strategy was working, for now there was little for the fire to feed on. In another hour the blaze was under control and eventually out altogether. Smouldering heather and grass, and an acre of gaunt, blackened spruce skeletons were the evidence of its passing. We gathered beside the smoking ruin of the car, faces and clothing filthy, exhausted but pleased with our prowess. I wondered what unpleasant truths Jimmie would level at its unfortunate owner.

A dishevelled Bob suddenly materialised from behind his vehicle. He was mopping his brow with a filthy handkerchief and looked both guilty and disconsolate.

'I've gone and lost the ignition key!' he announced, angrily.

It was too much. Suddenly, still shaking with delayed shock and utterly exhausted, we were all splitting our sides, hiccuping, spluttering, choking. I glanced at Jimmie and, once again, he had me beat. There was little there, now, to remind me of the frenzied maniac who had shinned up and down a tree in record time, raced like an Olympian over a rough hillside and fought a fire with unrelenting determination. Wig and hat restored, face miraculously clean, he had become his usual staid self. But, all at once, our boss was laughing, too. He had a sense of humour, after all.

Jimmie gave me a lift back to the junction for Coillessan and someone, to my huge relief, had thought to bring my bike from the Douglas Wood.

'Well done, lads,' he said to no one in particular, as I alighted. Then drove on with the others.

I was soon home. A bath, a meal, together with some small satisfaction that I had not made a fool of myself at the fire, set me up for the trip with the stalker in the evening. I was looking forward to it with mixed feelings knowing well why he was going to the den. To most sheep farmers the fox was bad news and any means to control its numbers was acceptable. Ardgartan was right in the middle of sheep country and local interests demanded that the forest, which was accused of harbouring the animals, must get rid of them. It was Alec's job to visit all the known dens each spring, discover which were being used, and then to shoot, or despatch by other means, both adults and cubs. I had grown up with this annual slaughter and for many years had taken it for granted. Now it was cause for concern. Reliable evidence against the fox seemed slim and, it seemed to me, much was simply prejudice handed down from one generation to another. I needed to know more of the animal and the problem concerning it. So, I set off, as arranged. Perhaps we could talk it out.

'I heard you had a spot of bother this afternoon,' Alec greeted me with a grin.

'Aye. It nearly got away.'

'Jimmie told me. Sure you want to come? We could put this off a day or two.'

'I'm fine. Let's go.'

Glen Croe was that long, narrow glen where Bill O'Connor lived. Alec had already noted wind direction for the approach to the den, so we drove to the top and parked his old van in a gateway. I took in, once again, an impressive landscape, steep ridges on either side clothed to the tree line with pine, larch and spruce, mountain tops behind, sheer rock faces, giant boulders, long scree slopes, and a sea of dark heather to tramp through.

'Whereabouts is it?' I asked, binoculars ready as we stood at the roadside.

'Look straight across to the ridge and the big rock fall at the bottom of the gully. It's in the middle of those big boulders.'

'Got it, I think. How close can we get?'

'Not too close. If the vixen's there and she discovers us, she'll move her family to another den as soon as we're away.'

We set off across the hillside, into the wind, a long trudge over a broken hillside and with almost no cover. If either adult saw us, no lack of our scent would help. Dog or vixen would disappear at once and it would be useless to wait or move to another position. They would not return. Because of the drought there was barely a trickle in the burn and we crossed it without difficulty. Then began a climb through rock and scree. A raven swept by on whispering wings, another one right behind – travellers in the dusk, winging home for the night.

'They're nesting on a crag in the next glen,' Alec whispered. 'There's an eagle's eyrie here.'

'Where, for goodness sake?'

'In the gully above the den. On a ledge high up.'

'Do you know if they are using it this year?'

'Oh, yes. She's been sitting a while now.'

'You might have told me.'

'I meant to. Sorry. Just forgot.'

I could have murdered him. Nesting in progress at an eagle's eyrie and the possibility of young to be seen later on. Just what I had hoped for. Well, there was still plenty of time for watching, I supposed.

In a little while, Alec called a halt.

'We'll stop here,' he announced and lifted his binoculars. 'See yon large rock standing by itself among a lot of smaller ones? It's covered in heather. The dark hole beneath is the entrance to the den.'

'Got it,' I said. 'What do you think? Is it being used?'

'Difficult to tell in this light,' replied the stalker. 'There wouldn't be much sign, anyway, until the cubs are old enough to come out.'

After a good look round with binoculars and discovering no foxes slinking through the heather creeping towards their den, we settled down for a long wait. Alec directed me to the eagle's eyrie and I noted a huge basket of sticks on a ledge high up in the fox den gully. I would certainly come back here, if it could be managed.

It was an incredible evening. Dusk falling fast and stars already appearing in a rapidly darkening sky. A full moon had just cleared the far ridge, but its light was not strong enough yet to compete with residual daylight. We waited without speaking. No wind at all, now. A great silence brooding over the glen, the only sound the faint rustlings of nearby creatures of the night and the sluggish gurgling of a burn. As time went on moonlight played havoc with imagination. Surely that was a fox beside the burn? Surely something was crawling through the rocks over there? And so on. My ears were strained for the slightest sound, eyes for the smallest secret movement.

But nothing happened. It became extremely cold and I wondered how long Alec would hang on. A small, lonely light beamed from a window in Bill O'Connor's cottage, a warm beacon in the immensity of the Highland night. I remembered enthusiastic applicants for the post of wife and thought, again, of the lowering mountains on either side of the glen, now mysterious in bright moonlight, and the tall, dark conifers clothing the lower slopes of the ridges. Sometimes, but rarely, the glen sparkled in sunshine, its burn tumbling prettily, silkily over a rocky bed down to the loch. More often, its magnificent scenery was only glimpsed through blankets of mist and the thundering of boiling waters, after a downpour, would drown out all thought. She would be a brave woman who was prepared to share this solitude.

Bill's light went out and immediately and as if on cue, the night was shattered by the piercing call of a fox. A harsh screeching , which echoed right around the glen, seemed to come from somewhere in the darkness below, perhaps not far from his cottage. We cringed, thinking the animal must discover us. Again the call came, a sharp imperative from one fox to another, a sound once heard by human ears that could never be forgotten. I shivered with excitement.

'Is that the vixen?' I whispered.

'I think so,' replied the stalker, and put a finger to his lips.

The possibilities raced through my mind. If it were her, surely she must have cubs in that den? Might we see both vixen and dog? I looked, once again, towards the dark shape beside me and Alec, guessing my thoughts, nodded agreement.

By now we were relying entirely on moonlight for vision. It was an extraordinary feeling standing there, all senses alert, knowing that somewhere, not too far away, a fox might be creeping over the hillside. I felt exposed, huge, unable to hide my bulk away, but the man beside me seemed calm enough and confident. He was absolutely still, his glasses focussed on the den.

There were no more calls, but, all at once, I felt Alec stiffen. Without moving, my heart thumping uncomfortably, I tried to copy the direction of his binoculars. And, spotted her! Not at the mouth of the den, as expected, but poised on a small boulder to one side. She was small. She was motionless. Against the sky I saw pricked ears, nose pointing, a twitching brush sweeping out behind, an animal waiting for something to happen. Greedily, I took in the dainty shape, silver red in the moonlight, and imagined her mate, somewhere in the darkness of the shadowed hill, making sure it was safe to bring her food. Her cubs would be curled in the warm den, waiting to snuggle back into her warm belly.

The dog did not come. I wondered, if perhaps from a different angle, he had our scent though the vixen was not worried. She still stood poised and alert. Then, in only a few more minutes, she leapt lightly down from her boulder and padded off down the hillside. I lost her almost at once as she slipped in and out of the rocks. Alec signalled that we should wait and perhaps he thought the dog, even now, would arrive with prey. In half-an-hour, though, cloud began to build over the eastern ridge and soon moonlight vanished. Time to go. We crept away, making a wide detour, and hoped that the pair were somewhere to windward of us and would remain unaware of our presence.

Back in the car and driving slowly home, I asked Alec his views on the necessity of controlling foxes.

'It's always been done,' he said, cautiously. I think he sensed disapproval.

'Is it really necessary?'

'Well, I've never seen a fox take a lamb but the shepherds say they do.'

'Do you have to do it?'

'Oh, yes. Jimmie has given me orders. He has to keep in with the local farmers.'

'A pity,' I sighed. 'I would have liked to watch this family.' There was a long silence as we coasted down the hill and in due course, arrived at the junction for Coillessan.

'Thanks, Alec,' I said, as I stood beside his open window. 'That was great. By the way, tomorrow I'm taking the launch down to Mark. Thought I'd take a look at the heronry and then try and find that badger sett. Somebody told me there were peregrines on the cliffs right at the end of the forest. Is that right?'

'Ah,' commented Alec, looking guilty. 'I meant to tell you there was a pair.'

Would this exasperating man never learn that I wanted to know about such things?

'I wondered,' I said, casually, 'whether you would like to come?'

'Aye, I would.' He seemed pleased.

'Good. Meet me at my house at seven. Bring something to eat. We'll probably be out all day.'

'Tell you what,' Alec said, as he fired the engine, 'If you like, I'll leave that den till the very last. You might see the cubs.'

seven: wild work

ALEC arrived punctually the next morning. The sky was overcast and a light rain was falling. Good for the newly planted trees on the hillside. I hoped it would clear later, for this was a long planned trip and whatever the weather, we were going to set out. Riding sluggishly to her anchor on a swell the *Mary Ann*, too, was ready to go, the boat we would use to board her drawn up on the shore.

'We'll stop off at the heronry,' I said, as we chugged away from Coillessan. 'The youngsters should be pretty active by now.'

'Good idea,' the stalker agreed. 'There's a nice omen for the day,' he added, pointing upwards.

A flock of whooper swans were dipping and rising in and out of the wispy cloud. I counted thirteen. Over the beat of our engine their haunting cries could not be heard but, for a moment, I imagined their wild calling and wished I could join them. Majestic white birds, great wings powering, slender necks ever reaching northward, they flew for old haunts on lonely islands and coasts. There, in a brief summer warmth, they would nest.

We sailed close to the shore. Patches of dirty white snow still clothed the mountain tops but, at sea level, spring had arrived at last. Ancient oak was bursting into leaf, birch was already well clad, ash, alder and rowan were not far behind and there was blossom on the geans. Once more that unhappy contrast struck me, the varied greens of deciduous trees, haphazardly growing wild where seed had fallen and taken root, against the solid mass of dark conifers planted in rigid formation behind. Okay, the conifers were being grown for commercial purposes and could be categorised as a crop to be harvested, but surely those straight lines, at least on the boundary edges, could be avoided and a mixture of species planted. Once again, I was dreaming of the perfect forests I would one day create – in cloud cuckoo land!

We passed the abandoned old croft house on the shore, watched the seals on their favourite rock and soon were approaching the aspen wood. The rain had ceased and a pale sun was breaking through the overcast. I cut the engine and for a few minutes, gently rocking, we watched to see what was going on. It all looked promising from a distance. Amongst silvery leaves all fluttering in the breeze, we could see chicks on the tops of their nests, restlessly treading the sticks as they waited for parents with food. Adults, big grey birds, ponderously purposeful, were flying to and from the heronry. Great.

'If we go ashore here,' suggested Alec, 'there's a good chance we won't be spotted.'

I started the engine and we gathered way, slowly, silently as possible, towards a pebble beach. The anchor was dropped and in minutes, hidden from the heronry by a rocky spit, we had rowed ashore. Over a rise I could see the aspen wood away to our left, but Alec led away from it over rather marshy ground towards some silver birches. A flash of brown and startling white was a sandpiper rising from somewhere close by. It called a trilling warning to its mate and flew swiftly for the shore, skimming the pebbles, darting this way and that, and finally alighting on a rock covered in wrack. In a few seconds it was joined by its mate and, as we turned away, both were bobbing up and down in great agitation calling their distress.

A few yards further on, we disturbed a totally unexpected bird. I had often seen small flocks of them in urgent flight over the loch, apparently on their way to mud-flats at the top. Good feeding there for waders. But here was a single specimen, a redshank hen sitting tight on her eggs. Startled by our sudden appearance she rose on thin, red legs from her nest, and flew with protesting cries for the open water, a dark brown bird with sharp, red bill, shimmering white on wings and tail. We stopped just long enough to admire her four buff, brown-spotted eggs snug in the grass-lined hollow, then hurriedly took ourselves off. I knew she would be back as soon as we were out of sight.

Beyond the birch wood, a plantation of mature spruce. A burn came trickling down the hillside from within it, its waters finding a sluggish way to the nearby loch – no break in the drought as yet. Alec made straight for it and, still moving fast, we followed its course.

'What's all the hurry?' I asked.

'Sharp eyes!' was all he said. But, of course, he was right. The heron seems to have eyes in the back of its head and is not easily stalked.

Very soon we were amongst tall conifers and the pace slackened. I heard the sound of flapping wings and through a gap in the canopy was just in time to see a large grey bird passing overhead – yellow dagger bill, head hunched grumpily into the neck, long legs trailing behind. It was an adult heron, perhaps bringing food to its young. There were no indignant squawks from above to indicate we had been detected and almost immediately, a great clamour broke out in the aspen wood as youngsters demanded to be fed.

The stalker had certainly been here before! All at once, he stopped beside a broad-stemmed tree. On the far side, a rope ladder hung from one of the thick lower branches and six-inch nails had been knocked into the trunk at vantage points on the way up.

'Of course, Jimmie will know all about this?' I teased, as I examined the rope for strength.

'He wouldn't mind.' Alec grinned.

'Let's get up there,' I said, and began a distinctly energetic ascent.

We sat on two branches, our backs against the stem, with a good view to the aspen wood through swaying boughs. Above our heads, the dense canopy would serve to hide us. Through my binoculars, I remembered the heronry at Eilean Righ – trees which all seemed to contain a nest and bespattered platforms of sticks seemingly precarious in their boughs. Here, youngsters, one, two, sometimes three, were standing stork-like, stiff as soldiers at attention. Some were dozing, some just gazing into space with staring, orange eyes that I knew would miss nothing that moved. Chicks began to preen – long white necks in a corkscrew twisting, sabre-sharp bills poking, poking at feathering uneven with down, puff-balls floating this way, that way on the breeze. Chicks began practising dangerous dances on the sticks – urgent wing-flapping to lift-off, nearly over the edge to an unknown world below, but safe on the nest again with a wobble. From time to time, a parent arrived with food. An incredible racket announced its imminent arrival. All the youngsters burst into song, their ratchet-like calling filling the air to the exclusion of all other sound. It only died away when the lucky one was being fed and the unlucky ones realised there was nothing doing. Silence then, till the next donor was seen flying in.

'Twelve nests?' I asked Alec, quietly.

'There's one or two more in the sitka,' he whispered. 'You can't see them from here.'

A crazy idea occurred. Wouldn't it be great to get nearer those chicks? Why not climb to one of their nests?

'Could we take a look?' I suggested. 'When all the adults are away?'

My companion understood at once. 'There's probably a sentinel bird keeping watch somewhere,' he said, doubtfully, but I could see he was tempted.

We had to work fast In a few moments we were scrambling down and making our way as quickly as possible towards the heronry. As we entered the aspen wood, I briefly noted the ground all carpeted in autumn's russet leaves and decorated, with the pale blue eggshell chippings of the present nesting-down wafted everywhere with the disturbance of our passing. Which would be the best trees? Which had a suitable nest? Looking anxiously upwards, it seemed that myriad silver-green leaves, all fluttering in the breeze, were excellent camouflage for the huge baskets of sticks which must be somewhere in their leafy wilderness. At last, relying on what we had seen from our perch in the sitka and somewhat desperate, we each chose a tree at random and started climbing.

I shinned up mine to the lowest branch, grabbed it and hauled myself over, and on. Where to go next? Above my head, another good branch easy to reach, and I was soon astride it. But then, a snag. I found myself much too close to the giant nest. It was immediately above my head, much larger than I had expected, and so beautifully constructed, the sticks so tightly knit, that there was no possibility of seeing through. At least there was total silence from the nest and the chicks, if chicks there were, were not bothered by the upheaval beneath them. Nevertheless, I began to panic. It suddenly seemed to be a long time since I'd started up. A shiver down my spine was expectation of immediate sentinel bird attention.

What to do? I'd better go down. A last glance all around and, just in time, I realised that the branches of the next tree intertwined with mine and that two of them were probably higher than the nest. Were they strong enough to bear my weight? Were there herons there? I would have to take a chance. I slid to the ground and then began to climb again as fast as possible. In no time at all I was almost level with the nest and all this time, dead silence. I began to expect it empty. With yet one more reaching upwards, I grabbed the branch I had thought suitable and with a mighty heave brought head and shoulders briefly above the rampart edge. In fleeting seconds, I took in two solemn youngsters standing on filthy, dropping-soiled sticks, skewer-bills pointing my way, reptilian eyes unblinkingly regarding me, and fluff balls of down dancing everywhere. The gangling chicks were almost completely feathered, well-feathered and so well-grown that I was afraid that, at any moment, the extra-ordinary apparition that was me would trigger off first flight. A dismayed youngster, somehow retrieved by Alec or myself, would stand little chance of being safely returned to its home. It would just be too difficult to get it there.

Time to go. I slid down most of the trunk, jumped the rest and took one more long look at the magnificent construction above, the first ever heron nest I had been close to. It was a mistake. Admiring its size and thinking of the cargo it held, I fielded an eyeful of squirted waste for my pains. A thump nearby was Alec coming to earth. He, too, had a filthy face.

'Any luck?' I asked.

'I couldn't get past the bloody nest,' he said in disgust, joining in the wiping-up operation.

Mark was our next objective. We hurried from the aspen wood straight to the shore, then over the spit into the little bay and, in only minutes were back in the *Mary Ann*. The sun had broken the cloud and visibility was good. In a loch that was utterly still, our wash rippled away behind and each small wavelet could be traced till it lapped the shore. The ridges and forests on either side were reflected in ever receding planes to as far as the eye could see. Perhaps it was too clear, too limpid and the bright spell would not last.

When we coasted into the bay at Mark, John's boat was missing. That almost certainly meant the old man was in good health. He would be over the loch visiting friends or doing some shopping.

We dropped anchor and rowed to the shore. The old cottage looked as lonely as ever. Smoke drifted from its chimney to tell us this was a lived-in place but the door was closed and curtains were drawn across each window. No welcome there. But, as we walked towards the steading, bedlam broke out from within a shed and then the collies came careering over the shore, wildly waving tails and joyous barking, to greet us. We had been recognised and were welcome.

'I'll need to take a look at his sheep before we start searching for badgers,' I said.

'I doubt that'll take long.' Alec grinned.

And he was right. Accompanied by the troupe, who miraculously seemed to understand Alec's orders and remained at our heels, we went first to the fields where they should have been. Not a single woolly creature in sight! We made straight for the wood and the new plantation and there they all were, happily nibbling at succulent spring shoots on the infant trees. One or two heads were lifted, incurious eyes assessed us, then feeding was resumed. The dogs showed no interest and were obviously not used to shepherding the creatures away from their illicit feeding grounds. Something would have to be done and I did not think Old John would like it.

'Right,' I said. 'We'll have a decco at the field where the badgers have been feeding and take it from there.'

A stroll to the first of the fields revealed the fencing material, brought on our last trip, still rolled and stacked where it had been left. The

tiresome old bugger! In the other field, we found the young spring grass
dying in patches, as before, and some quite large areas brown and bare
to the soil. Larvae could still be seen.

'Okay,' I said. 'Let's see if we can find any badger runs.'

We spent some minutes surveying the hillside. Away to our left was
the cliff I had climbed to rescue Cronk the raven. Its rocky ledges fell
more or less straight into the loch, so no chance of a badger sett there.
To our right, above John's bay, was the wood where the sheep were
placidly feeding. We had seen no signs of badger use in it. We decided
to begin beside the inescapable proof that they were around, the
droppings in the top field, and made our way to the wall which enclosed
it. If badgers were coming for beetle larvae, then they would tend to
follow a familiar path in. And, they were. A narrow, padded down and
obviously much used path came from the birches higher up the hill and
led straight to the old wall. Here it was unbroken, its weathered stones
still in place. Were the animals going over the top? They were not. In the
middle of a clump of nettles we found a cunning tunnel dug beneath,
and on the other side an exit concealed by another large patch.

Following the track was easy at first, but once among the birches
there was a surprise. It did not continue into the wood but suddenly bore
left, parallel with the loch and across more open ground. It did not seem
likely there could be a sett so close to the shore, but we continued along,
threading in and out of scattered birch, losing the path sometimes in
patches of dead bracken but always picking it up again on the other
side. Eventually, we arrived at the large burn which was Old John's
water supply. It chuckled a mocking message: 'no badgers here', as it
bubbled merrily past. I agreed with it! Better to have been halfway up
the hillside by now.

And then, another surprise. The track vanished into the burn. We
followed it on to the bank, down between some rocks and then into the
water and it did not emerge on the other side. Badgers need to drink, of
course, but something interested me. This track looked so regularly
used. Did they come for a drink at this spot every night? It was possible
but seemed unlikely. Later on in my career, I would have read the signs
at once, looked for prints, looked for droppings and correctly interpreted

their messages. For now, I was puzzled. We retraced our steps and discovered a few yards back, in a confusion of bracken and bramble branches, that the track had in fact divided. One to the burn. One climbing to the wood.

'Might as well see what's on the bank further down,' I suggested.

'Right,' agreed Alec, I thought a trifle reluctantly but he did not explain.

There were other small paths lower down which disappeared into the burn. Narrow tracks also led towards a broad marshy area close to the loch. We came to a flattened down patch of grass with remains of crab and the head of what I thought was an eel on it. To one side was a dark dropping which smelt of fish and had small bones bound up in it. At last, alert for prints, I found them on a nearby patch of damp soil – five toes on each foot, claws well dug in, and best proof of all, webbing between the toes clearly marked. Of course!

'It seems our badger's an otter!' I laughed. 'There might be a holt somewhere close.'

Alec seemed only moderately interested and I could see he was afraid we were going to go looking for otter holts.

'Okay,' I said. 'What's bugging you?'

'Need to get up the hill,' was all he would say.

We rejoined the original track and now began to climb on a gentle rise. About 200 yards from the shore, among scattered birches, we came to a large outcrop with lots of loose boulders on sloping ground beneath. And there it was. Our track came to an end in a cavity under a large rock. Not a badger hole, though. Beside it, on ground blackened and damp, was a small dropping. It smelt of fish and contained the now familiar splinters of bone. This was an otter holt.

'Did you know there were otters down here?' I asked, pleased with our find.

No,' replied the stalker, looking slightly embarrassed. 'My job's deer.'

That was one way of looking at it, I supposed, but was surprised. Surely he could not fail to note the wild life that daily he must see about him.

'Better get on,' said Alec, gruffly.

A track of some sort, otter, badger, roe deer, led on from above the outcrop and into the wood. Now we were climbing through birch, oak and ash and making for the larch plantation through which I had first walked down to Mark. Somewhere on the way to it, we lost our trail. Then we came to a ride which divided larch from spruce. It was steep and rocky but pleasant with spring grass, deepest green moss in damper places, and primroses just coming into flower. Alec led the way and seemed anxious to get wherever we were going as fast as possible. Fitter than I, he tore up that hill as if life depended on it and, of course, he knew the ground far better. But, toiling along behind, I suddenly remembered the cliff beneath which I had encountered the wild cattle. Surely, the rocks at its base were a likely place for a badger sett? Okay, there had been no obvious tracks in the ride, other than those of deer, and no give-away droppings but as we came parallel with the place, some instinct prompted a look. I called a warning to my companion and hurried into the larch wood.

Giant boulders had fallen from the cliffs and settled on the hillside. They had been there long before the majestic trees, but like ancient monuments they now stood guardians of their cathedral place. Stately stems towered to a ceiling of gently stirring branches and russet needles provided a velvet covering to the floor. Amongst them, plain for anyone to see, I found the badger sett. Three impressive platforms of dug-out soil, impacted over many years and covered in scratch marks, gave away the position of three huge holes each beneath an enormous rock. Undoubtedly badger holes. How could I have missed this? Narrow, padded tracks connected each of the holes and also led on into the wood. On one of these, close beside a rock covered in moss, I found a heap of badger droppings in a scraped out hollow. There were signs, also, of much activity – needles scraped aside by industrious paws, old bracken stems crushed and broken, banks down which bodies had slithered and slid. I wondered if there were cubs.

'Well, I'm damned!' exclaimed Alec, when he responded to my call.

'Surely, you knew about these?' I teased.

'I've a large area to cover,' he responded, looking rather offended.

'Only joking,' I said. 'If you hadn't been in such a hurry, we might never have come this way.'

If Alec was annoyed, he must soon have worked off his spleen by continuing his punishing pace up the hill. He almost ran and I had a job keeping up with this mountain goat. But after an hour, when we had reached the end of planted trees and had crossed a fair bit of open hillside towards a higher ridge, he finally called a halt. We sat on some heather in the lea of a large rock.

'What's all the hurry?' I panted.

'There's a long way to go, yet,' was the cheerful response, 'and I don't think the weather'll hold.'

He could be right. A bank of heavy cloud seemed to be building far away to the west. As we ate our sandwiches, I looked around with interest. To the east the ground rose gradually, perhaps over limestone rock, first to lush grassland on which sheep were grazing, and then to heather moorland. In the far distance I could see our new forest area and the seemingly endless ploughed land still to be planted. Right above us, the ridge. I knew from my map that on the other side of it steeply sloping banks of grass and heather came to an end in cliffs which fell sheer to the sea. They were the most westerly part of our peninsula and marked, also, the limit to Ardgartan Forest. Right in front of us, a small loch. Alec told me it was called the Corran Lochan. Its still, dark waters reflected every detail of the sombre cliff behind and its brooding depths seemed somehow full of mystery, perhaps with the secrets of many a dastardly deed in the past! Present reality was a small flock of mallard who sailed forth from reeds at the far end and sent ripples chasing each other to the shore. Two gulls made plaintive music as they planed overhead and a pair of ravens, talking to each other, flew from gathering cloud. No doubt, they noted our presence below.

The cliffs were magnificent. Above those calm waters and beyond an area of cotton grass, they rose step upon step, ledge upon ledge of granite rock, to a height of about 800 feet. Here and there, brilliant outcrops of sparkling quartz relieved their gloom and on a few ledges, small rowan saplings were clinging to life. Two large gullies broke the steep sides, one immediately in front of us, the other at the east end of the loch. A mountain burn trickled down through each. Both would be impressive after heavy rainfall, but now their waters wandered un-obtrusively through rocky courses and were lost in pebbles on the shore.

We were sheltered from the wind and it was pleasant to have a break. I was jolted out of thoughts of badgers and otters when Alec touched my arm.

'See yon gully,' he said, pointing to the first of the gullies. 'Left side, close to a rowan about halfway up. There's a buzzard on a rock.'

While I searched in vain, the commentary continued.

'There's a ledge behind the rowan. The nest may be there.'

At last I found him, a dark bird against shadowed cliff, perched motionless on a knife edge. So well-camouflaged was he that, surely, only movement could have given him away. I watched his head turning this way and that, pausing when he must have spotted us and was sure he would fly but, for the moment he stayed, a sharp-eyed sentinel surveying his world.

'How did you find him?' I asked.

'I see the pair every year and know where to look' replied the stalker. 'They always nest up there. The keeper on Corran is on at me, every year, to destroy the nest.'

'Good grief. Why?'

'Says there are too many buzzards and they take his grouse.'

'That's nonsense. There are plenty of rabbits.'

'Well, that's what he thinks. He'd poison the lot if he could.'

I sensed disapproval and was surprised. The attitudes of gamekeepers and stalkers to predatory birds were nothing new to me. Why should Alec be concerned?

'Does that bother you?' I asked.

'I suppose so. They're beautiful creatures.'

There was hope, yet, for the man!

All of a sudden our buzzard, definitely a male, lifted off his perch and began flying towards the moor. It was purposeful flight, broad wings rhythmically flapping, the bird never deviating from his course. In a few minutes he arrived over the heather and began, at once, to quarter the area. Head down, gimlet eyes searching, he hunted for prey. A tiny movement below must have caught his attention. Gliding smoothly to the spot he hung on the air to watch and wait. He lost height, flapped madly to regain it, circled round to the same position and continued his absorbed contemplation. Was there a vole down there? Would he pounce? But, no. Evidently there was nothing doing. The bird fluttered away towards another patch of heather, rose into the breeze, and then again was inspecting the ground below.

Hurry up, I thought. Catch something. Take it back to your mate – certain confirmation of a nesting situation. But he caught nothing and I wondered whether there was a shortage of voles that year. Then, a diversion. Something caused the bird to veer away abruptly from his hunting. Now speed was of the essence. He soared towards the clouds, gaining height all the while, and going like the wind. Higher and higher he rose into the breeze all the time making towards the cliff where the nest was. What was the matter?

'Hey. Look at that!' Alec exclaimed.

Out of the clouds beyond the ridge, a tiny speck was hurtling towards our buzzard, a winged bullet on its way to a target. 'Peregrine!' I declared, not daring to take my eyes from the bird.

'Yep. I think there's going to be fun.'

The falcon streaked towards the buzzard, passed below, looped effortlessly upwards to a height, then came swooping down towards it. At the last second it swerved away to avoid collision – quick flick of long, slender wings and tail, and perfect timing. The buzzard wobbled in his steady flight but otherwise took no particular notice or avoiding action. He just continued straight on. Twice more the attacker came plummeting in, once from below and then from above, each time it seemed missing the object of his fury by inches. But the buzzard flew on serenely, apparently disdainful of this impertinent intruder.

'What's going on?' I asked.

'It's territorial! Buzzard nesting on this side of the ridge, peregrine eyrie on the other – on the cliffs.'

If you could call it a battle, it was soon over. There seemed to be a buffer zone above the ridge which both birds observed as neutral. Perhaps certain 'messages' were exchanged – buzzard: keep off my patch, falcon: don't come near mine. Who knows? Anyway, the buzzard suddenly veered away and began flying back towards his hunting grounds and the peregrine, with a cheeky dip of its wings, soared swiftly into the clouds and vanished.

'Better move,' said the stalker, now the action was over.

He yanked his bag into place and began walking as fast as was possible through the cotton grass and towards the loch. Now, its waters were ruffled by a small breeze and the cloud was heavy over the sea. Rain was coming. Heather lined the loch on this side and we turned along a narrow path which was probably used by fishermen. The mallard ducks scurried away from the bank as we passed, uttering deep quacks of disapproval. They probably had nests nearby but we did not stop to look. We came to the reeds at the end of the lochan. Two hoodie crows rose with raucous cries from something lying there. The ground was wet and treacherously boggy. Treading delicately over heaving peat, we came to some planks and an old gate laid across mud and water. A dead ewe lay caught in one of its bars. She had either drowned or starved, and now stank powerfully – food for hoodies and maybe buzzards too, if prey was difficult to find.

Once over the bog we made for the big gully to our left and began to climb. Dead bracken, broken by single birches, gradually gave way to heather patches and rough rock scattered on the steep slope. As the gully narrowed towards the top, its sides grew ever steeper and bare of vegetation. We followed one bank of the burn and this provided convenient steps up the hillside. Climbing steadily, not talking but saving our breath, we heard a sudden, plaintive call ring out from somewhere on the rocks to our left: 'pee-ou, pee-ou'. The call was distinctive and repeated again and again. Certainly, a buzzard. But the male was still hunting over the moor, so this had to be his mate. Then a big brown bird flew right above our heads. Larger than the other and with a pale necklet on the speckled feathering of its breast, it was undoubtedly the female. She circled above us, all the while uttering cries of warning or distress. Had she come off eggs? Immediately, I was anxious, for these were nervous birds which would desert after only small disturbance from human beings.

'What do you think?' I asked Alec. 'Should we get away?'

He looked up and now the bird was flying steadily towards the moor, perhaps to join the other.

'It should be all right,' he said. 'If we're quick.'

I was not quite sure what the stalker was up to but the next ten minutes were painful – rocks with jutting jagged edges, smooth rock with no holds of any sort, bunches of heather grabbed when nothing else offered, a handful of holly bush, patches of wet and treacherous moss. When we finally came to a halt, I thought we must be close to the buzzard's ledge.

'Stay there,' Alec whispered.

I held on to a nearby rock and sat down – it seemed the safest thing to do in this precarious place – and watched the man climbing towards a smooth ledge which had some vegetation growing on it. Then he was crawling along it. A buttress appeared to block his way. A rowan grew on the far side. The rowan of the buzzard nest? Now the stalker was on his stomach and wriggling forward. Then he was peering over the edge and down to something below. Moments later, and apparently satisfied, he began to reverse and in seconds was crouching, once more, beside me.

'Hurry,' he said. 'Go and have a look.'

No time for questions. I edged past him, still apprehensive, and repeated his action. In a few moments I, too, was in a position from which I could look down. And there, indeed, was the nest of our buzzard. The site was well-chosen. The untidy bundle of sticks was beautifully secure, protected from wind and sun by a shelf of rock which gained added shelter from the rowan sapling. In the hollow of its centre, which was lined with twigs and grasses, lay three large, whitish eggs marked with red-brown blotches.

There came renewed wistful calling from over the moor and, at the same time, a hiss of warning from Alec. In seconds I was back beside him.

'Let's go,' I threw at him, still sure that our hen would desert.

'She'll be okay,' he assured me quietly, but I noticed he led away directly and was moving as fast as was possible.

In order to avoid further upsetting the birds, we scrambled and slid as quickly as we could straight down the cliff and into the gully. Alec reckoned they would see us go and thus be reassured. We crossed the burn and began to move up the gentler slope on the other side. Near the top, we made for a large boulder standing in the middle of a patch of heather. We crawled into its shelter and hoped long, straggling branches would conceal us from sharp buzzard eyes. On the other side of the gully, far below, we could just see the ledge of their nest.

The stalker smiled. 'Don't worry,' he said. 'I know this bird. She'll go back.'

I hoped he was right and settled with binoculars at the ready. The pair continued to circle round and round over the moor, not calling now, not hunting. Endless minutes passed. Then the hen suddenly made up her mind. She broke away from her mate and came flapping back across the gully. From high above the ridge, she came gliding down towards the cliff, then alighted on that knife-edge of rock from which the cock had first risen. She stayed only a moment. I watched her look all around, crane her neck to check her nest then float softly down to alight on it. She did not rise again but settled down over her eggs and I could relax.

We crawled the few yards to the top of the gully then, once over on the other side of the ridge, paused for a rest.

'That was pretty risky,' I suggested, referring to the visit to the nest.

'I thought you'd like to see it.'

'Hm. Better later on when she's hatched and well-bonded to her chicks.'

'She's never deserted yet and I usually have a look each spring.'

'Okay,' I smiled, and left it at that.

Cloud was now thickening overhead and I was not surprised when Alec set off again at a smart pace. There was no more climbing to do. Instead we walked down over undulating ground, peat bog which was dried out with the drought and through heather patches on long stretches of spring grass heavily cropped by sheep. And then we were there, right on the top of the cliffs and looking out to sea. There was a wonderful feeling of space and I stood breathing in salt air appreciatively. The sea, however, leaden in the poor light, looked a hell

of a long way down and I hoped there were no hair-raising balancing
acts ahead of us.

The stalker led down the side of a narrow cleft. I followed, watching
scree rolling away from our boots and disappearing into the void – there
did not seem too much to hang on to in a tight moment. I thought Alec
was going to show me the nest of one of the thousands of seabirds flying
over the cliffs and wondered why the man was in such a hurry. Then we
were climbing the other side towards a ledge. Arriving there, I found it
reasonably wide. Alec eased off his rucksack and then sat down with his
back to the cliff. Apparently unconcerned by the abyss beneath, he
pulled out his pipe and lighted it.

'There you are,' he said with satisfaction. 'They haven't spotted us.'

'Who hasn't?' I asked, mystified.

'The peregrines!'

So that was it, and I knew what he meant. Long ago, at Kilmartin, I
had heard the clamour from these birds when they find an intruder in
their territory. It was surprising that so far we had not been discovered
for I knew they must now be nesting. Once again, I wondered if Alec
really knew what he was doing.

'Where's the nest?'

'It's below that overhang,' he replied, pointing downwards and to the
right. 'You can't actually see the ledge.'

This was a marvellous place. The cliffs plunged in steep, rock steps to
the sea and at no point could I spot where they actually met the water.
The air was filled with the wild calling of seabirds, some already busy
with their nesting, others staking out small territories for whenever it
would begin. I caught glimpses of each as they flew urgently out to sea
and back again to the ledges below, gulls, cormorants, guillemots and
razor bills. The sea came rolling in on a mounting swell to surge against
the rocks below, irresistible, unstoppable power, then withdrew in a
hissing wall of water to gather itself for the next assault.

I knew peregrines to be much bolder birds than buzzards so was not
too worried about disturbing them. They would be further on with the
breeding process, too, so were firmly bonded to their nest. As long as we
took ourselves off shortly after they had discovered our presence, there

was little danger of a desertion. I settled back to enjoy whatever might come and did not have too long to wait. 'Kek, kek, kek', came the call from somewhere along the cliff, sudden, angry, blotting out all seabird sounds. A dark, streamlined bird flashed past, swooping sharply up towards the sky then disappeared behind the ridge. Alec grinned and made a thumbs up sign. The bird streaked past again, long, grey tapering wings and a glimpse of a speculative eye. 'Kek, kek, kek' it scolded, the call fading as it passed out of sight.

'The cock,' he noted, laconically.

The bird came flying past once more and settled briefly on a pinnacle of rock some way to our left – just time to take in the slate-grey back, dark moustache, fierce yellow-rimmed eyes, and hooked bill. He shrieked dismay and took off in a graceful curving, right wing dipping, tail ruddering, and vanished once more behind the cliff. Not for long. This time I picked him out high above us, hovering on fluttering wings, head down observing, but silent. Still silent, he flew away out of sight and I wondered if he had given up.

Not at all. His mate had flown to join him. Now the peregrine pair began a superb display of aerobatics, mostly filling the air with their noisy shrieking, but occasionally so silent it was almost eerie. Sometimes working together, most often in tandem, first one, then the other, unbelievably fast, incredibly graceful, they kept hurtling towards us and swerving away at the last possible moment. They stooped, they looped the loop, they changed course, they soared skywards only to come diving down again, and always they came from behind the ridge so that the first we knew of their coming was the sound of their raucous scolding. I sat enthralled by the sheer skill and speed of their action, the poetry of perfect motion.

'Better go, now,' said Alec, suddenly. 'They've had enough.'

He led the way back into the cleft, the way we had come, and began to climb its crumbling sides as quickly as possible. The peregrine pair, perhaps sensing our departure, were now circling above our heads. Silent, wheeling in wide, sweeping circles, they watched to see what we would do. By the time we had reached the top, only one bird was still there. I hoped the hen was back on her eggs.

Now, I noticed a large cairn, over to our right, which someone had built on the edge of the cliff. Surely it was right above the peregrine eyrie? Alec was making straight for it and, as we drew nearer, intrigued, I saw that it was covered in droppings. Perhaps the tiercel used it as a look-out post.

'Why would anyone want to mark this spot?' I asked the stalker. 'It's not a summit.'

'You'll see,' he replied, and added grimly, 'I expect.'

We were there in seconds and I looked curiously at this landmark which seemed to hold such significance for my companion. 'Look on the side above the cliff,' he urged.

There it was, a rusty length of chain hanging from the heavy stone which marked the top of the cairn. Attached was a gin trap, evil and ready set with a dead grouse for bait. I could not believe it. This was forest land and neither Jimmie nor I would countenance such a barbaric device used on it.

'Who put that there?' I asked, angrily.

'It'll be the keeper from Corran again,' Alec replied, apparently not a bit surpised. 'He thinks peregrines, as well as buzzards, take all his grouse!'

'They probably take some,' I agreed, 'but he's no business to do this without permission.'

I was horrified at the thought of a bird caught by the leg, struggling to escape, and eventually dying from its injuries or starvation.

'Does he ever catch one this way?'

'Oh, yes. It's usually a juvenile. When they're first out of the nest, there's always a chance. I found one, dead, last year.'

'I'll speak to Jimmie,' I said.

Alec wrenched the contraption away from the cairn and, with a grand gesture, threw the lot over the cliff.

'That'll teach the bugger,' he remarked, with satisfaction.

'Check as often as you can,' I said. 'Let me know if another is set.'

Before the war peregrines had been mercilessly persecuted by game-keepers on account of their supposedly heavy predation upon grouse, and in pigeon fancying areas, further south, they were at risk because they killed racing pigeons. Peregrines were also shot during the war because it was thought that pigeons were carrying highly confidential messages from one secret destination to another and should be protected. Apart from a belief that the bird did not do the amount of damage attributed to it and an innate dislike for the cruelty of gin traps, I knew that its numbers had seriously declined and was damned if Ardgartan Forest was going to help it along to extinction.

As only it could in this part of the world, cloud suddenly enveloped us in a thick, drenching mist. One minute, we could see for hundreds of yards to the top of the ridge, the next, it was white-out and we were enveloped in a silence in which there were no peregrines, no screaming seabirds, no birds at all. Better get a move on for Mark and the *Mary Ann*. I turned right for the ridge and the gully that would take us down to the Corran Lochan.

'Stop,' yelled Alec. 'Do you want us both in the sea?'

Looking down, I got a shock. The edge of the cliff was only feet away. A bad moment.

'Thanks,' I said. 'I was sure we came that way. You'd better lead.'

The trip down was uneventful in spite of the mist – Alec could have made it in pitch darkness, I reckoned. There was nothing to hold us up at Mark, either, for John was not yet home. Back in the boat, tired and neither of us speaking much, I made plans to come again if time could be found. What with badgers and otters, both perhaps with cubs, and the buzzard and falcon pairs, there was plenty of action to follow up. Then, as we sailed into the bay at Coillessan, Alec dropped a bombshell.

'I'll be up in the glen on Monday,' he said. 'Maybe I'll see those fox cubs.'

'Don't you dare go near that den,' I protested, then noticed a sardonic twinkle in his eyes.

'Only teasing,' he replied happily, thus reciprocating my shaft of the morning in regard to otters and badgers.

We solemnly shook hands on it. Nevertheless, the question remained. Did he really just think of these animals as vermin to be dealt with each spring? In spite of his concern that I should see the cubs, I had a horrid feeling that he did.

eight: cattle calamities

I NEVER did get to see more of the badgers and otters at Mark. A visit there was rare and because the rain which had engulfed us on the peregrine cliff only lasted overnight and we quickly returned to drought conditions, the risk of fire continued. As many men as I could spare were patrolling the forest during working hours checking for fires – one area was particularly vulnerable because many walkers came that way to the hill and a carelessly dropped cigarette would be enough to start a major conflagration amongst young trees. Evenings and weekends were also covered, a roster being drawn up of available manpower. I did not catch Jimmie's spring fever, but I was never off duty and began to pray for rain heavy and prolonged. Jimmie, himself, was also cause for concern. The man was either not well or was feeling his years – retirement could not be too far off. As neurotic as ever about checking for fires with his telescope, he nevertheless left the affairs of the forest entirely in my hands. Not that I minded, but now and again responsibility hung heavy. It was I who had to decide what jobs needed doing and which were most urgent, who directed our now considerable workforce, who checked on progress and generally kept an eye on a huge forest area. No van was available for use in the forest, so I had to walk or bike everywhere. This was healthy but time-consuming. There was certainly no time for watching wildlife.

But I did get a break. One morning, at dawn, my phone rang and the caller reported what might be a small fire on the other side of the pass at the top of Glen Croe. No car and a long, hurried slog on my bike up the hill ended in the discovery of a modest camp fire, rather damp and smoky, on which a camping couple were cooking an early breakfast. After briefly speaking with them I remembered it was still early. Why not take a look at that fox den in the glen which Alec had shown me? After friendly parting warnings about starting a forest fire, I free-wheeled down to the place where we had previously left his van and set off across the hillside. From the side of the road, binoculars at once picked out the gully and the scattering of boulders where it was situated. That bit was easy, but I was much too far away to see what signs there were of use, whether cubs might be playing there, or even the exact position of the hole. Must get much nearer.

The wind was good, not much of it and blowing across the glen from west to east and therefore in my face. I started through the heather wondering if I could follow the same route as before. Once again, I admired the magnificent scenery. The sky was clear, the sun not far from rising over the eastern ridge, and a golden light enveloped the landscape adding warmth to the dark conifers on either side. Larch was in bud and its fresh green, pale against pine and spruce, confirmed that a conifer forest could be pleasing to look at provided the species were mixed to provide contrast in colour and shape.

The light steadily improved and deer tracks took me in the right direction. I soon found the huge boulder where we had previously stood. Binoculars balanced on its top, I prepared to wait for at least an hour. After that it would be unlikely for foxes to be out and about. I waited for precisely five minutes. It was not a fox, though, that provided the action. Not at first. A strange call suddenly rang out across the glen, a sharp yelping sound rather like the impatient yapping of a collie dog gathering recalcitrant sheep. But there was no-one gathering sheep and, in fact, no sheep at that moment in the glen. The sound was repeated and I realised it must come from somewhere on the opposite hillside. At the same moment and not connecting the two at all, I remembered the eagle eyrie Alec had pointed out and which I had never found time to come back and watch. The ledge with its huge construction of sticks was easy to find, and that solved the problem. A large bird was perched there right on the edge. Then I saw an impressive hooked bill open and close and, seconds later, heard that same weird call. An eagle calling to its mate?

The great bird lifted off, floating from its eyrie on long dark wings. The sun caught its golden head as it glided smoothly down and highlighted the warm, brown feathering on back and wings. I lost it against dark conifers and then, startled, picked it up again plummeting towards the boulders at the bottom of the gully. It came like a rocket and I thought it must crash into the heather, but with amazing control for such a large bird, it veered away at the last moment and began steadily climbing for the sky. It began to circle the glen planing round and round overhead with only occasional flaps of its wings to maintain height. Then, once again, it was stooping towards the earth. Towards exactly the same spot, I thought.

This time, I discovered why. As the bird once again banked gracefully away, I spotted a splash of rufous red. A fox! It sat, back against a rock, watching its tormentor rising into the sky. The dark cavern which must be the den was a few feet away. The animal was small but larger than a cub would be at this time in the season. It had to be the vixen we had seen before. Fascinated, I watched psychological warfare. The eagle kept stooping down, deadly accurate, an arrow direct to its target, but each time veered away just above the small animal. Once, the vixen threw a vicious paw at her tormentor, but mostly, she just crouched and snarled as the predator came. Not knowing at this stage much about eagles, I wondered if one could possibly kill an adult fox and thought it unlikely, except in exceptional circumstances. So what was this magnificent bird after? My heart missed a beat. Were the cubs already coming out of the den to play? Much later on in my career I knew that it might have been, but for now I could see there were none.

The eagle tired eventually. It flew low down the glen, flapped slowly over to the east ridge then a few minutes later, from a considerable height and right over my head, was apparently making straight for the eyrie. Steadily, and purposefully, the bird dropped lower and lower: I lost it. Found it. Saw it lift to its ledge and drop. Soon it settled down in the giant stick nest. Then, only a minute or two later and just picked out against the dark cliff, another eagle arrived from further up the glen. It was smaller than the first, as is mostly the case with the males of raptor species, so I reckoned it was the hen who had done battle with a vixen. Perhaps her strident calls earlier had been a sharp imperative, for the newcomer had brought in a hare.

The vixen did not hang about. The action over, she ran straight for her den. There was time now to note the ground all around not greatly padded down and there seemed to be just the one track leading to it – definitely no cubs out yet playing rough-and-tumble games. Their mother would mostly stay with them, only emerging to collect prey from her mate and perhaps to rest from an increasingly demanding family. She would not go far. I planned to come back again in a week or two to watch for the cubs. Perhaps there would be further interaction between eagles with chicks hatched and hungry, and a vixen with youngsters she could not for ever protect.

It was not to be. A few days later, shortly after I had returned to my house after work, Alec came to my door. He looked upset and his hand kept flicking at his red moustache, a sure sign that something was wrong.

'It's Currach,' he stuttered. 'And I've buggered up your den.'

'What's happened?'

'I'm sorry,' he said, 'I was up Croe this morning. Thought I'd look at the den from where we watched before. I had Currach with me. She must have picked up a scent and was away before I could stop her. She made straight for it, of course.'

'Did you not call her?' I asked angrily. This was certain disaster for my fox watching plans.

'She's usually so obedient,' he said, unhappily.

'So what's the problem, now?' I enquired. I was ready to give the man a rocket for his carelessness – he had no business to be there with a dog at all – but I realised there was more to come.

'She's missing. I saw her right in front of the den. She probably went in. I walked over and called her. Nothing. I've been there all day.'

That was the beginning of the great terrier rescue that failed. The stalker was fond of his grizzled old friend and would not rest until he knew what had happened to her. He thought she must still be down the hole and somehow stuck, for otherwise, if free, she would have found her own way home longsince. The next morning, I lent him a few men from the nursery squad and told him to go and see what could be done. Three men with spades accompanied him, each man prepared to dig all day if necessary. He came to my house in the evening.

'No luck,' he said, sadly. 'We dug a long way in, came to a rock and couldn't get beyond. I thought I heard a yap but we've been at it all day and no result. I reckon she's dead.'

'Look,' I said, sympathetically. 'I can't spare anyone tomorrow but if you want to try again, that's okay.'

He looked greatly relieved and I hoped he would be successful. Much as I disliked the purpose for which these dogs were used, namely the putting to flight of the vixen so that, as she emerged, she could be shot, and then the killing of the cubs within the den, I had no wish for one of them to die a lingering death by starvation. Alas, Alec failed to reach the little dog, heard not a sound from down in the hole, and decided regretfully that even if she was still alive there was nothing more to be done. Currach had had it.

A week later, after work, I was home and having a word with the ponies in their stable when Alec arrived in his van. He was beaming.

'Look here,' he called through the open window.

Currach, sound asleep, bedraggled and very thin was lying on the seat beside him. She woke, briefly, to wag her tail for me and then collapsed once more.

'What happened?'

'Postie picked her up on the Croe road. She was making for home.'

We decided that the terrier had been so far down in the fox's den that, having killed all the cubs and eaten them, she had grown too fat to squeeze past the large boulder which had stymied the diggers. A whole week had gone by before, nothing more to eat and starving, she was thin enough to make it out again.

The drought came to an end in the middle of May, first of all with a tremendous thunderstorm and torrential rain, when all the burns on the hillside burst their banks and the earth drank greedily, then with showery weather interspersed with bright intervals. The young trees in the new plantation began to look as if they wanted to live and we started planting again. I was able to order the ending of regular watering in the nursery – a supply from the nearby loch through pipes was possible but labour intensive. In any case, the seasonal work there was finished for the time being. A stand of Norway spruce was ready for thinning so a squad, together with two of the ponies, was sent up there. A lot of sitka spruce needed brashing, too, so as soon as I was able to transfer the women from the nursery, they went up to this plantation with their ganger, Bill O'Connor.

It looked like we were nicely back to normal, but very soon there was another knotty problem to sort. One afternoon, I stopped at my house to collect some papers for Jimmie Reid. I found my kitchen invaded. Bill O'Connor and Katy were sitting at the table each with a mug of tea to hand. My door was never locked and, within limits, anyone was free to make use of the kitchen, but this was not the time for a break and both should be at work out on the hill.

'What's going on?' I asked.

'Katy's been attacked!' declared the Irishman, making it sound as though she had, at very least, been raped – he had a lively imagination. 'I thought it best to bring her here.'

'What's happened, Katy?' I turned to her, afraid that it was something really serious. She looked shaken and her freckled face was pale. 'Have you been hurt?'

'I'm fine, now, Don,' she replied. 'One of the wild cows went for me. It was my fault, really.'

'We'd nearly come to the end of a section, yesterday,' Bill intervened, 'and were due to start on a new one today. It didn't seem worthwhile taking the whole squad just to finish off a few trees, so I asked Katy to do them and then to join us later.'

The man appeared embarrassed, as if he thought he might have done the wrong thing. I'd better hear what Katy had to say. 'Tell me what happened,' I asked the girl.

She assured me that she had been pleased with Bill's suggestion. She enjoyed working on her own and as she tramped quietly up the hill from

Coillessan, unaccompanied by the usual babble from her companions, had thought she might see deer, or even a fox. She was disappointed, but as she turned off the forest road to walk down the ride to the sitka wood, she hoped there was still a chance.

There was, but not quite as she expected. To arrive at the sitka she had to pass through larch already thinned. Larch woods are lovely and this one particularly so, with sunshine pouring through the canopy, branches gently swaying and crisp needles on grass softly responding to each footfall. As she trod among them, the hushed silence was broken only by the swish of her feet, a joyful robin perched on the branches of a holly bush and a tiny wren scolding from beneath some brushwood. She was standing for a moment to enjoy the atmosphere of the place when, suddenly, a dreadful moan rent the air.

'It was horrible,' said Katy. 'Like someone had hurt themselves badly. And it seemed to come from somewhere quite close.'

The girl had stood rooted to the spot not certain whether to fly or whether to try and discover the cause. When another lugubrious groaning occurred she made up her mind. The cry seemed almost human. Perhaps someone was in need of help. When the next heart-rending ululation rent the air she made cautiously towards it. Eerie silence followed but she continued, hoping for the best. She came to the edge of a small clearing, a place of lush grass with a single old oak in the middle of it and a scattered birch or two in the heather – a nice little spot which some long ago forester had not had the heart to plant up with conifers. On the far side, a compartment of young Norway spruce provided the backdrop to an interesting scene.

'Do you know who my groaner was, Don?' she enquired, wide-eyed and obviously enjoying the telling of her tale. 'It was one of the wild cattle beasts. She was standing in front of the spruce and at her feet was a calf. She had just given birth.'

The girl had quietly withdrawn without being discovered and then had stood, a few yards back, against the substantial stem of a larch. From there she watched a charming scene. The cow stood panting. The afterbirth was expelled and then, as with most mammals wild or domestic, she nibbled and ate a little. Katy thought she was a young animal, but the mother certainly knew what she must do. She stretched her long neck to the dark object at her feet and, rough tongue at work, licked at the slippery envelope encasing it, nudging the little creature, urging it into movement and life. Soon it was trying to stand, wobbling on uncertain legs, subsiding, trying again, succeeding. Gangling and gawky, it swayed drunkenly on shaking limbs, just managing to remain upright while the mother continued her energetic cleaning-up. The calf was a heifer. Gradually a most beautiful creature with a thick and curly coat was revealed, coffee-coloured whilst wet, but when dry a delicate honey. Her head was daintily feminine, her eyes deep brown with long, silky lashes and her coat like velvet gleaming in the rays of the morning sun. Soon, with unerring instinct, she was nosing along her parent's flank, searching for the bulging udder.

'I should have gone, then,' admitted Katy. 'I knew perfectly well how fierce the cows can be when they have calves. I just got carried away and wanted to see more.'

Without thinking, she began to tip-toe nearer and accidentally trod on a brittle stick. Crack! Cleaning-up operations ceased at once. Up came the cow's head. Searching eyes immediately spotted Katy. There was a loud bellow then, head lowered, heels kicking and tail lashing, calf rudely knocked to the ground, she came galloping towards her. Unable to move, unable even to think, instinct took the girl leaping backwards, her brashing saw ready for action. Knowing better than to turn and run, she stood with a tree at her back to see what the crazy animal would do. The cow shuddered to a halt only yards away then, motionless, regarded the intruder with angry eyes.

Katy remained very, very still. The cow, seemingly reassured, trotted back to her calf. Now was the moment for the girl to quietly walk away, but she hesitated too long. Though the mother's head went down to check that her infant was safe, it was up again immediately and turned towards Katy. Once more on its feet, the calf began to search for milk. But the cow was still nervous. Could not settle. Kept eyeing the girl warily.

Suddenly, she butted the youngster away, tossed her head angrily and began circling around it, first at an agitated walk and then beginning to trot, the circles becoming ever wider and closer to the girl. It looked like at any moment she would turn and, head down, come charging.

At last Katy took action. The next time the angry mother had turned away, she took a few steps backwards to the first available tree and froze. Then, at each opportunity to the next and the next. At last, behind a thick trunk some distance from the clearing, she stopped just long enough to see the wild cow allow her calf to stagger to its feet and find her udder, then fled for home.

Katy had obviously enjoyed telling her story and it had probably received some embellishment as she went along. Nevertheless, she had certainly had a fright and been in some danger.

'But, you're all right?' I enquired. 'No damage done?'

'I'm fine, Don, but the rest of the women are nervous. It's not the first time a cow has looked like getting nasty.'

At this moment Alec arrived in the yard with a roe deer carcass tied to the back of Mollie. While Katy ran out to greet her beloved pony, I told him of the girl's experience and asked about the wild cattle situation at that time.

'The cows are calving around now,' the stalker replied. 'They can be really dangerous when they have young. Also, there are far too many bulls because none of the male calves are castrated. When the cows are in season they become very aggressive. Fights are common.'

'What about the owner?' I asked. 'They're his responsibility.'

'Jimmie has spoken to him many times. Nothing's ever done.'

'I'll have a word with him. How would you deal with the problem, Alec?'

'Me?' His eyebrows shot up in mock surprise. 'I'd castrate the lot!'

One day, not too long after this episode, after a morning checking a wood of Norway spruce, I returned home to pick up some lunch. I thought I heard voices as I came out of the Coillessan Wood and as I approached the yard I saw Alec standing by the gate. He signalled me to come quietly and I thought he must have found some injured creature which he thought I would like to care for. Once beside him, I discovered another man present in the yard, Douglas Fergusson, the vet from the village. He was behaving in a most peculiar fashion. A scalpel in his hands, he seemed to be playing Grandmother's Footsteps, the stable door the objective he must reach. From within came the sound of thudding hooves.

'Is one of the ponies ill?' I asked Alec quietly, wondering why the vet needed a knife.

'Er, no. Not exactly.'

Before he could explain, and I thinking we had pony problems of some sort, Douglas cautiously opened the creaking door of the stable. An amazing sight met my eyes. A huge red bull was tossing his head frantically and stamping thunderous hooves on the stone floor. One of the wild bulls of the forest! I could see a rope thrown over his neck which was secured to a ring on the far wall. At the sound of rusty hinges creaking, the shaggy head swung round and, for a second, wild eyes stared. Briefly, the creature stood still. Then he went beserk.

'What the hell's going on?' I shouted above the incredible uproar which followed.

'He was here when I came with oats for the ponies,' replied Alec, at the top of his voice. 'I sent for Douglas. He's going to castrate him!'

This was not an adequate explanation, but that could come later. More important was what to do now. For a fatal moment I hesitated. The creature was secured. The vet was here. It would be one bull the less to father wild calves. Why not let him go ahead? But the animal was dangerous. There might be a nasty accident. I dithered and was too late.

'I'll fetch a leg halter,' yelled Alex above the confusion, dashing over to the shed where the ponies' tack was kept.

The intrepid vet was nearing the door of the stall.

'Come back, Douglas,' I bawled. 'It's too risky,'

Our voices triggered off a tremendous roar from the bull and the sound of splintering wood, but the vet had not heard me. Surely the idiot would see it was impossible. He did not. He threw open the door and in spite of the plunging hooves, grabbed the tail of the frantic beast. Then he raised his knife to do its business.

Beyond idiot man and plunging beast, I caught a horrifying glimpse of a ring on the wall loosened and coming away.

'Douglas, look out, the ring's going,' I screamed.

It came away. The bull realized freedom, at once. He rose majestically on hind legs, pivoted round with scrabbling forefeet on the shattered wall, then with lowered head galloped out of the stable. Just in time the vet, with prodigious effort, threw himself over the partition into the next stall. As the bull came thundering out, I dodged left and was lucky. Alex, returning to the yard with the halter, received no warning at all and found the maddened animal coming straight for him. The bull saw an object he could vent his rage on. He roared and with head down once more, made towards the defenceless ranger. He lifted him high on his horns, scooped him up helplessly as a child, then tossed him shouting for help, right into a heap of manure by the wall. Then he bolted out of the yard and galloped, gloriously free, towards the Coillessan wood. The Waterfall Burn was cleared in magnificent style, a flash of gleaming red coat, head held high and heels kicking, then he disappeared among the trees.

Luckily, Douglas was not hurt. He stood brushing himself down and grinning.

'Are you all right?' I asked Alec, a trifle anxiously. He could have been badly mauled, but now stood holding his nose and trying to brush some of the filth from his breeches.

'I'm fine,' he replied sheepishly, knowing he was in for a ticking off.

'Serves you right,' laughed Douglas, as we hosed down the stinking fellow. 'Fancy cutting a grand beast like that!' Then he went on his way to another appointment.

'You and I have things to talk about,' I said to Alec, as we watched him go. 'Jimmie will not be pleased.'

'I know,' he said apologetically, shivering in his wet clothes. 'I don't know what came over me.'

'I'll lend you some dry gear,' I said, 'and we'll talk over a cup of tea.'

'It seemed too good a chance to miss,' Alec continued later. 'That beast must have smelt the hay in the barn and come looking. He followed his nose into the stall and was having a good feed when I arrived. I've got you, I thought, thinking of Katy's adventure and before I quite knew what I was doing, I'd slammed the door on him and thrown a rope over his neck. I used your phone and Douglas arrived within the half hour.

Kilmartin and the Knapdale hills to the south.

Caledon woodland in hard winter frost.

Sheltered by a rickety hessian screen, the women's team
transferred seedlings to the transplanting boards.

A team of men are pegging boards into the transplant lines.

The tractor driver halted in order to level up the transplant line.

In the nursery, skilled work from the tractor driver
placing soil against the transplant roots.

Birch in the early morning.

Forests are great places for pony trekking .

Planting trees – 'we only had to plant a thousand in a day!'

Forest plantations were fenced to keep out the deer.

Mist and mystery in the forest.

Ponies and men were great working partners.

There was great teamwork between horsemen and ponies.

Unhitching logs from the swingle tree – a pony working in the forest.

Lonely birch on Loch Awe side.

The next day I had a meeting with Jimmie and yet again he surprised me. That normally grim man, whom nobody could think of as having a sense of humour, nearly split his sides over the account of the attempted castration of a bull and seemed to overlook the small matter of his authority having not been sought.

'I've just heard that the Local Authority has notified old Bash Blair that they must be T.B. tested,' he said. 'Now, he'll have to bring them in, won't he?'

The farmer, with a final dire warning, did in fact make every effort to round up his cattle using his own rather meagre manpower and resources. But they proved so wild and dangerous he found it almost impossible to hire men who were prepared to handle the job. On one occasion a special corral was built on the far side of the peninsula, a number of bulls were successfully rounded up and after much effort, persuaded in. By the following morning the whole corral was flat on the ground and all the animals gone. Another few were driven, with difficulty, into a cattle float and the gate securely closed behind them. Within a few minutes the floor was in splinters and the legs of the panicked creatures were through and reaching for the ground. Two courageous volunteers managed to open the gate again but by the time they were all out, the float was a complete write-off.

Shortly after these unsuccessful attempts, Jimmie sent for me.

'Bash Blair has phoned,' he began. 'He agrees that the only course is to shoot the bulls. Get something organised with Alec, will you. Use men from the squads if necessary.'

'What about the cows and calves?' I asked, envisaging a massive and distasteful slaughter.

'He thinks they'll be easily gathered once the bulls are out of the way.'

We made a plan. The stalker suggested that there were few, if any, animals in the older part of the forest, so a drive west down the peninsula from Coillessan should work. He would lay on as many shooters as he could and deploy them on the hillside at Mark. Forest workers would be used to drive the beasts towards them. Carcasses would be butchered on the spot, into pieces that could be manhandled or carried by the ponies down to the lochside, and starting from Mark, the launch would pick them up at certain points along the shore, then take them up to the village.

The day duly arrived and everyone assembled in the stable yard at Coillessan. Alec noted the direction of a light wind with satisfaction. From the north west, it soon broke-up the lingering mist and the men, who were driving, would have it in their faces and the cattle not scent them as they came. I was introduced to four game-keepers, all from local estates and recently discharged from the Army, and two shepherds who had come over from farms in the next glen – all new faces to me. Crack shots, they would be using .303 rifles for the job. Gangers Andy Murray and Bill O'Connor were in charge of the squads who were delighted to have a break from normal duties and Peter and Dick, who were busy grooming Jock and Tommie, would bring the ponies through the forest later on, to help carry the carcasses down to the loch. All in all, it was a big operation which, if unsuccessful, would have used a great deal of time and manpower.

'Right,' I said, at last, when everyone was ready. 'You all know what this is about. Alec's in charge, though I'll be around for an hour or two. The shooters will come with us in the launch down to Mark. Andy and Bill are staying here with the rest of you. You'll be spread out in a line to cover the hill and when they give the signal, keep in position and move towards Mark. We'll be ready. The wind is good. Keep it in your faces. One thing more. These are dangerous beasts. I want no heroics. Be careful. Good luck all.'

We chugged quietly down the loch, the water mirror-like, our wash a small ripple spreading to infinity. There was no time to admire the beauties of nature, however, for Andy and Bill would be giving us precisely an hour and a half before starting to drive. Alec rehearsed plans with the shooters and I sat at the helm, going through the motions but secretly regretting the action that was to follow. Wild they might be, but these bulls were magnificent beasts, proud and free and after all only behaving as nature would prompt them. However, with more and more people beginning to walk in the forest for recreation, I knew they had to go. It was a shame.

Forty minutes later, still on a calm, unruffled surface, we slipped almost silently into the little bay at Mark. The mist had not lifted yet and the old grey cottage looked ethereal, even ghostly, in a backdrop of drifting cloud. There were no collies lying at the door. There was no sign of life. Old John must still be a-bed.

'I'll have a quick word with him,' I said to Alec, when we were all ashore. 'He doesn't know about this, yet. Get your lads organised and I'll join you shortly.'

I would not be leaving the lochside so there was plenty of time in hand. I made a roundabout approach to the steading through the new plantation. It was as it had been before. No change. Ruminant eyes regarded me with indifference and woolly heads went down to nibble. They were everywhere. I was disappointed but not surprised. The old man either could not afford to feed his flock, or he was just being bolshie and defying orders. But, I had no choice, now. Jimmie would have to be told and, doubtless, action of some sort would be taken.

A bleary-eyed John opened the door as I came through the garden and the collies, shut up in the shed but hearing me at last, burst into song.

'Morning John. How are you?'

'Fine,' he replied, but I thought he looked rough.

'We've a shoot on today. The cattle. There are six marksmen down here with Alec, so don't be surprised when you hear shots. Stay at home.'

'I heard about it,' he replied, grumpily, 'at the pub.'

'The sheep?' I enquired, no elaboration necessary.

The old man exploded. 'You can tell Jimmie Reid that if he sends anyone to check on me again, I'll take a candle to the forest!'

Alec had once told me of this oft-repeated threat, usually made in the pub by Old John when he was grumbling about the latest interference from Authority. He had thought it all out. A lighted candle placed beneath a small tree in the plantation would gradually burn down and set alight the dry litter on the ground. He would have long since disappeared over the loch to buy provisions. When the resulting great fire got going, no one would connect it with him and there'd be no evidence left either. Or so he reckoned.

I smiled. 'You know you don't mean that, John. You've told that tale too often and you'd never get away with it.'

He grinned but did not contest the point. I said cheerio and promised to be down again soon.

Alec was standing alone by the dinghy, the rest making their way to whatever positions they thought strategically good. After an agreed interval they would start moving slowly towards Coillessan.

'Let's go.'

'Not far.' He chuckled. 'We'll wait for a while in the wood across the bay. The cattle are often there.'

This, I remembered, was where I had had an adventure with a bull jammed between two rocks. It seemed a long time ago. A pleasant spot it was, with patches of grass and heather dotted with rock in widely spaced oak and birch. No bracken invasion as yet. We stood with a small outcrop at our backs beside a heather-covered boulder. This would give us cover from behind and Alec a broad field of fire if a bull came our way. An hour went by. There was silence from up the hill. No shots, no shouting from men driving cattle. Still much too early to think something had happened to abort the operation, I nevertheless began to hope that, perhaps, no beasts would come our way.

Another half hour had passed when Alec suddenly nudged my arm and pointed. At first I could not see what he was after, but I heard munching jaws and sucking sounds. To our left, partially hidden by the outcrop, a wild cow stood quietly chewing cud and beside her a calf, quite large by now, was noisily drinking from her udder. They had come silently, had not scented us, and now stood contentedly in a patch of grass completely unaware of our presence. Both looked in excellent condition.

'Hell,' swore Alec under his breath. 'If a bull comes this way, we may have to shoot her, too. She'll go mad.'

All at once the full implications of this operation hit me. The bulls were usually to be found grazing in a loose group more or less together, but in the breeding season they separated each to defend his own small harem. If any mothers were shot then their progeny, left orphaned, would have to killed as well or somehow captured and removed to the farm. It seemed like there might be wholesale slaughter and I wanted no part of it.

'I'll go now,' I whispered to Alec, not wishing to witness this event. 'I need to look at the Norway plantation near the heronry. Might as well carry on from there to Coillessan. I'll keep to the shore. See me back at the house.'

He grinned and I knew I was kidding myself that he did not understand my need to leave. He said nothing, however, just gave a thumbs up acknowledgment. I managed to make use of the outcrop as cover and think I got away without disturbing the cow.

Down on the shore, the tide was ebbing. It was peaceful enough, the day now basking in sunshine, but I was edgy, waiting for something to happen in the forest. When, at last, three shots rang out, short, sharp spittings of death that cracked the silence, I noted them not too far up the hill and just about parallel with my position. I imagined a group of panicked bulls charging through the woods but whether one, two or three had been shot I could have no idea. None of the animals came my way. After that, the shots were intermittent and I lost count of them.

I walked up to the Norway plantation and found a lot of damage to the young trees – deer damage. Leaders and side shoots had been well and truly eaten. Perhaps the stalker had missed some in the last cull or maybe the enclosure fence had somewhere been breached. I spent an hour walking round it and discovered a large gap in a culvert through which an enterprising hungry animal might squeeze. Others would follow. Alec would have to come back to deal with them.

I did not go as far as the heronry for by now, all the youngsters would have fledged and be flying around in the wood. I found the spruce which Alec and I had climbed, its rope ladder still intact, and climbed up to the branch we had sat on. Nothing much was happening in the nearby aspen wood. One or two immature birds were flying around rather aimlessly, never going far, always returning to base. Others stood on the old nests preening or just sleeping. I reckoned a good few had already discovered the loch and were probably fishing. Of parents, there was no sign. Half an hour went by and I was beginning to think that it was time to go when I became aware of a soft scratching sound, a rustling in the dead and dry foliage that littered the forest floor. It seemed to come from below me and sounded quite close. A mouse? Surely not. There was no way at this height that I could hear such a small creature. Energetic movement continued. Could this possibly be a badger burrowing for worms? Well, if it was, it was one with unusually poor scenting powers and a strange liking for being out in the middle of the day!

I spent some anxious moments trying to manipulate thick sitka branches silently out of the way. Momentarily, the rustlings ceased and I held my breath, but they began again, a continuous and busy working away at whatever it was doing. At last, lying flat on my belly along a branch, I managed a peephole through the needles. And nearly laughed out loud. At the base of a tree, only yards away from mine, a dainty eider duck was completely absorbed in the important business of building her nest. Although well up the hillside and away from the loch, she had evidently decided that here, right in amongst the spruce trees, was the best and most convenient place. Perhaps she had always nested there.

For the next ten minutes I was the fascinated spectator of an elemental occasion in the life of an eider duck. With clumsy webbed feet, she continued scraping out a sizeable hollow. Then she began tramping round and round it, drawing in twigs and needles to form a small barricade between herself and the forest floor. She crooned softly to herself as she worked, as if content with the job she was doing. A comfortable nest took shape. She cautiously tried it out for size, settling her plump body down into the hollow and drawing in more dead vegetation about her until her sides were nearly covered. It was done. Madam was satisfied. She stood again, checked carefully all around for intruders, then with a solemn and preoccupied air, as if something enormously important was about to happen, put her head down and her tail up. Out popped a large, smooth green egg. My lady turned to regard it complacently then, with broad bill once more at work, edged it into position beneath the soft feathers of her breast and settled down. Her eyes closed.

It was a nice little episode, but how was I to escape? Between four and six eggs would probably be laid. I did not want to wait that long! Luckily, there was no problem. In a few minutes the eider duck was up on her feet. She industriously drew soft down from her breast to cover the egg, added a few twigs for good measure, then waddled off down the hill. Laying would be completed later. I gave her ten minutes then slid down from my viewing point.

Alas, there was not a happy ending to the story. I returned a few weeks later to see how she was doing. There was no sign of the lovingly constructed nest, nor even of a scraping on the forest floor. Had my eider never completed her clutch? Had some predator stolen her eggs? I would never know.

Meanwhile, shaking a shower of needles out of my clothing, l walked down to the shore. There had been no shots for at least an hour and I had thought I heard the *Mary Ann* chugging up the loch – maybe with the first of her loads. I'd better hurry home. Alec would be reporting.

As I passed through the wood at Coillessan I heard the launch sail into the bay. The stalker had seen me coming.

'We got ten bulls,' he told me. 'No cows, you'll be glad to hear I've just taken the shooters up to the village. We'll need at least four more trips to ferry carcasses.'

'Well done. What about the boys?'

'I fixed up for the lorry to meet them on the forest road. They're on their way, now. Peter and Dick will walk the ponies back or see if Old John can put them up overnight.'

'Good,' I said. 'Get along home as soon as you can and see me tomorrow.'

The next day we discussed the situation. Alec reckoned there were certainly ten more bulls still roaming the forest. Another big operation was out of the question and, by and large, this one had not been as successful as we had hoped. He suggested that he begin culling the rest in the same way as he would be culling the deer, namely stalking them and shooting them one by one. I agreed and told him to start as soon as possible.

'You did well, yesterday,' I congratulated him. 'It was a tough job. You look buggered. Take the day off.'

'I'm fine,' he muttered gruffly. But he looked pleased.

It took all of three months before we got the wild cattle operation completed. Alec shot twelve more bulls, making a total of twenty-two and afterwards, on a happier note, the cows soon settled down and were gathered with little difficulty back to the home farm. They quickly became domesticated.

nine: a sheepish exit

SEPTEMBER arrived and I still had not made the promised visit to
Mark. With Jimmie apparently house or office bound these days, forest
affairs occupied all my time. One evening, however, Dougal came to speak
to me at home. The old woodsman had been there the previous weekend
to see his friend, John. He was worried. The man was not well, he
thought, and was not looking after himself properly. Also, he had not been
across the loch for several weeks and neighbours on the other side, with
whom he usually had a crack, had not seen him for a while. His sheep
were all over the place, as usual, and were not doing the trees any good.
Old Dougal shrugged his shoulders and wondered if something should be
done.

It was obviously high time I went to see the old man. He might be ill
or too infirm to remain on the lonely croft, or he might simply be thinking
he could get away with murder as far as the sheep were concerned. I
decided to go down as soon as possible and this time Jimmie must come,
too, in case there were awkward decisions to be made. Somewhat to my
surprise, when I put the matter to him, he agreed and a few days later,
after breakfast when the squads had already departed to various jobs, he
arrived in his old banger. He gave me a gruff 'good morning' and I
noticed with amusement that he had brought his telescope along. We
rowed out to the *Mary Ann* and on a glassy sea set out over the loch. That

he had nothing to say was normal for this time of the day, but I thought he was more morose than usual, as if something weighty was on his mind.

'The old chap's been with us a long time,' he volunteered, a last, 'but either he or the sheep will have to go. I can't put up with the damage they are doing any longer.'

'He'll take it badly.'

'Yes.'

He did not elaborate and we sailed on without chatting. It was a nice crisp day, not a breath of wind, and with more than a hint of autumn in the air. Once, I saw the old chap glancing up the hillside, to the Douglas Wood. Thinning had long been completed but I thought I detected approval in his eyes. He said nothing, but long after he had retired, I knew it would be a monument to his careful planning and a job well done. Shortly after, we were passing the old derelict croft on the shore and it was sad to think that Old John's cottage might, one day, be in the same condition. If he decided to leave, it was unlikely that anyone else would want to live in such a remote location. We came to the aspen wood with its heronry. I thought of that journey seemingly a long time ago, when Jimmie had sent me on foot to Mark and I had become hopelessly lost in the sitka plantation close-by. He said nothing, but he gave me a wicked grin. The young herons had long since departed, but now and again we passed one on the shore, stiff-legged, stiff-necked and severe, waiting patiently to catch a meal. Aspen, oak and birch leaves were turning to autumn yellow, and patches of snow touched the summit ridges on either side of the loch. Both were reminders that in a month or so my first year at Ardgartan would be over.

In due course, we arrived off the little bay at Mark. I cut the engine and we drifted slowly in with the tide, all the while searching for signs of life. Old John should have been at his work, but neither he nor the collies were anywhere to be seen. No smoke curled from the crumbling chimney, significant in a part of the world where the fire was seldom allowed to die, and the door, usually open to wind and weather was closed. Was something wrong? There was an aura of brooding desolation over the place and I even imagined the old fellow gone from there, the cottage beginning its inevitable decline into ruin.

Jimmie was kneeling in the bottom of the launch, his elbows propped on the bench and his 'scope' balanced on the side. From this uncomfortable angle, with the boat rocking gently, he was trying to assess the sheep damage to the plantation. At last he sighed, packed it away and made ready to transfer to the dinghy.

'I'll need to take a look,' he said, when, in a few minutes, we were securing its rope to a post. 'John's not in the plantation – probably still in his bed. Tell him I'm here. That'll give him a chance to sort himself.'

'Right,' I replied. 'He must have heard the launch. I'll give him a few minutes.'

I gave John ten, then walked across the shore to the garden. All was unusually quiet and though the door to the shed was open, the dogs were nowhere to be seen. Then, all at once, the usual cacophony broke out. It came from within the cottage. Why were they there? Hurriedly, I knocked on the door. Nothing happened, except more canine hysteria. I shouted. Still no response from John. Could he be somewhere outside? It seemed unlikely for he would have released the collies and surely come to meet us. I hesitated. Tried the door, which I knew would never be locked and stepped in.

The rabble tore past my legs without the usual welcome and raced for the shore. I entered the kitchen and stood in the dim light from partially curtained windows. It seemed worse than usual. Complete and utter confusion. Two kitchen chairs had been over-turned and one had a leg broken. A rug that usually lay in front of the stove was now in a crumpled heap beneath the table. The settee, much used by the dogs, had been turned to face the wall and was shoved against it. I wondered why. There was a pile of dog shit in one corner made, I knew, by animals who had become desperate. There were the usual odours, only much worse – stale cooking, stale tobacco smoke, a strong smell of whisky, and unwashed pots and crockery in the sink. Surprising was a plate on the table with a half-eaten meal on it – John liked his nosh and even unadulterated baked beans from a tin were a popular repast. It was the familiar pile on the table, however, that really brought home that something was wrong. There remained just a few packets and a couple of tins of corned beef. What was wrong, and where was the old reprobate?

I was just about to foray up the rickety staircase to the bedroom when a stentorian snort from the far side of the room put my fears to rest. At least, he was alive. The sound seemed to come from the settee and when I looked over its tall back, there he was, fully clothed, eyes closed, face flushed, his feet on one of the substantial arms, his head on the other. An empty whisky bottle rose and fell on his chest and from beneath a blanket askew came gentle snoring. Undoubtedly, this was a giant hangover.

I was not particularly surprised. Forest workers, like Old John, often had to live in remote areas far from other habitation. Often they were not married or had wives who, no longer able to tolerate the lonely life, had long since left. Each day, after work, they returned to an empty, un-welcoming home, had to prepare the main meal of the day, and then get through the rest of the evening with only assorted dogs, cats and sometimes other strange pets for company. In winter, evenings were long and, of course, Highland weather could be awful. No television in those days and radio reception often impossible, there was no other entertainment, no easy transport to visit friends. Boredom and depression were common and the whisky bottle the only solace.

It all depended on the man, of course. Many used their free time to pursue various hobbies. Carving from wood or horn was done with great skill and minute attention to detail – models of ships, animals and birds, even domestic pets were painstakingly constructed. Many were except-ionally well-read, borrowing books from the nearest library, sending away for something fancied from a catalogue, and even buying one or two on rare visits to town. Some beautiful gardens were created with loving care, and flowerbeds and rockeries often blazed with colour for a greater part of the year. Many grew vegetables, for the local shops rarely stocked more than potatoes, carrots, cabbages and onions and even they were not usually fresh. Lawns were patches of green perfection where daisies, buttercups and moss were never allowed to gain a hold.

For some this was a perfectly acceptable way of life but for Old John, perhaps, it had become too difficult and lonely. Either he was not well or he no longer cared. I was wondering how to get him sorted before Jimmie arrived, when there was an extra loud snort and a fit of coughing. His eyes slowly opened and then he was regarding me with none too friendly a stare.

'Are you all right, John?' I asked, genuinely anxious.

'What do you mean, am I all right?' he enquired thickly. 'What do you want?'

'Jimmie Reid is here. He's come about your sheep. He'll be in shortly. Better tidy up.'

'Oh, aye?' His eyes closed again and his head flopped down on a cushion. 'Why can't you leave me alone?' he groaned. 'They don't do any harm.'

Who's he kidding, I wondered. John made no move to sit up or to leave his improvised bed and I noticed his breathing was rapid and laboured. Suddenly, I began to be concerned. Perhaps he was not just the worse for drink?

'Are you sure you're all right, John?'

He sat up slowly, rubbing his face with grubby hands. 'I'm fine.'

Then the whisky on his breath hit me full in the face. A hangover, after all! Well, I had to somehow get him sobered up so that when he met Jimmie he would understand the purpose of the visit. I pulled the couch away from the wall.

'Look, John. Get your head in a basin of water. Have a wash. Make us all some tea. I'm away to fetch Jimmie.'

He seemed to get the message. Moaning quietly, perhaps for my benefit, I couldn't tell, he put his feet to the floor then, defeated, put his head into his hands and remained still. I thought I caught him peeping through his fingers, assessing the effect of this performance, but guessed that he would now pull himself together. I hurried away.

The forester was on the shore staring glumly out to sea. 'Is he all right?' he enquired.

'Been on the bottle.'

'It's bad,' he continued, referring to the sheep damage in the plantation. 'There's no way I can let things go on as they are.'

'He's drunk,' I said. 'But I don't think the old fellow's fit, either.'

'Maybe.'

There was a long pause during which both of us were probably mulling over the problem. The tide rippled at our feet and mocked our concern.

'Right,' announced Jimmie at last, with unusual firmness. 'Let's get this over.'

At the cottage, we found the old man, apparently restored, clearing a space on the table for three mugs. The kettle was singing on the hob and he was muttering away under his breath, no doubt lambasting interfering Authority. I thought he looked a little better. Jimmie's eyes swept quickly round the little room and when he spoke it was gently but firmly, and to the point.

'The fences have not been mended, John. Your sheep are still in the plantation. They will have to go, or you will. I'm sorry. Which is it to be?'

There was stillness in the little room, a pregnant silence during which each of us, no doubt, was conscious that a point of no return had been reached. Jimmie, expressionless, stood by the door waiting. I, unable to keep still, wandered over to one of the windows. John sat at the table staring at its dirty, streaked top.

'I'm getting on a bit,' he volunteered, at last. 'Maybe the sheep are getting too much for me. I'd better sell them.'

'I think that's wise, John,' said Jimmie, quietly. 'Do you know anyone who'd like to buy them?'

'Oh, yes,' he surprised us. 'Old Cruickshank'll have them. He has a farm near Lanark.'

'We'll get something fixed up for you,' offered the boss, looking mildly amused. The old fellow had obviously already planned that far ahead. 'They can be ferried across to Finnart in the launch. Could Mr Cruickshank send a float?'

'Aye. He'll do that.'

'Right,' said Jimmie, briskly. 'That's that. Are you fit, John?'

'Just had a touch of the flu,' he replied, sublimely sure his words would be believed.

'Take care of yourself, then. I'll send Alec down with some tucker to keep you going till you're better.'

A laconic hand was raised from the table, but John said nothing more. We drank a quick cup of tea with him, called in the dogs, fed them, then closed the door behind us.

As we strolled across the shore, Jimmie was chuckling. 'I think the old bugger had got it all fixed, don't you?'

'Reckon, so,' I laughed. 'He wanted some free transport.'

Ferrying sheep across a loch was an unusual assignment for a forester, but there it was. Jimmie had ordered it so. Having got his part in the business over, he gave me the task of organising the trip and putting it into effect. There were probably forty sheep in all. The *Mary Ann*, on a calm day, would take twenty. So two trips across the loch would be needed. The tricky bit was the tide and the available time within which to work. High water and the hour or so afterwards would be crucial, partly so that the distance between the gathering field and the launch would be as short as possible, and partly to enable the *Mary Ann* to be run in close to the shore and held on her painter – it would be far too difficult and time consuming to use the dinghy and row, say two animals at a time, out to it. No men could be spared from the squads but I would take Alec with me – he was well used to handling sheep of his own and would be invaluable. With John and his dogs to help, we ought to be able to manage. I had it all neatly planned and though I did not look forward to parting the old man from his unruly flock, I reckoned it would work. Or would it?

The fateful day arrived and, appropriately as it turned out, it was Friday 13th October. The loch was as calm as we could possibly have hoped for but to match our mood and the occasion, it was enveloped in a thick blanket of mist, suffocating and depressing. Everything was dripping – tears from heaven for an unhappy old man? Alec joined me at Coillessan and we set off straight away. The trip down was uneventful, the *Mary Ann* carving a silent passage through still waters and in and out of swirling curtains of mist. Only the mournful cries of gulls speeded us on our way. At Mark we could hear yapping and yelping from the collies and a great bleating of sheep. Good. Though we could see nothing, John had apparently gathered his flock into the field, ready for us to take over. He must have come to terms with their imminent departure. I looked anxiously at the sky. We could not afford to wait for the mist to clear, yet the job would be difficult if it did not.

The launch slid with a gentle crunch on to a pebble ramp, a passage to the water painstakingly devised by a previous inhabitant of the croft. Alec leapt over the side and took the painter ashore to the fence stob always used for this purpose. He wound it once around then brought the end back to the boat – this way it could be let out with the ebbing tide and the *Mary Ann* remain afloat. I was securing that end to a hook in the bow, when the sun broke through the cloud. Exactly on time!

A scene of total confusion met our eyes. No wonder we could hear the dogs. Sheep were running everywhere, over the shore, into the woods, up the hill, into the plantation. None was safely contained within the field. The collies were there, black and white bodies, yapping and snapping, darting in and out of the blackfaced, woolly-coated, bleating animals who were all fast becoming thoroughly bemused and had not a clue as to what they were supposed to do. Only one of the dogs knew his business, an older fellow, but each time he rounded up a few of the animals, his enthusiastic mates came tearing in to help and scattered them again. It was bedlam. Of John, there was no sign.

'Christ,' exclaimed Alec. 'The old bugger!'

'Better try the house, first,' I suggested. 'See what's going on.'

Conscious that time and tide were of the essence, we ran over the shore through a melee of distracted sheep and dogs. We hurried through the garden and when we came to the door an old ewe walked sedately out, as if she owned the place. Later, we were to recognise her as the leader of the flock. Old John was sitting in a chair beside the stove, staring morosely at the floor. A bottle of whisky stood at his feet and he was holding a half-emptied tumbler to his lips. He seemed unconscious of the noise outside and the confusion it represented.

'You've come then.' He stated the obvious and seemed cheerful enough. 'They're ready for you in the field.'

'They are not, John,' I said indignantly. 'They're all over the place. It's too bad. Why are you not watching them?'

'I was just having a dram to wish them luck,' he replied, unashamedly. 'They must have knocked the gate over.'

I'll knock you over, I muttered to myself. Alec had wandered outside again. I could hear him calling the collies.

'Come on John,' I insisted, 'Give us a hand. We'll use the dogs. But, hurry.'

We did not use the dogs, or at least four out of five of them.

'They're useless,' reported Alec, when I joined him. 'Old Ben's okay, but we'll have to shut the others away. They're not trained at all.'

This we did with difficulty, but at last the unruly mob was safely confined in the shed and we could get to work. The old dog had once been a champion and now, no longer confused by his exuberant companions, he did well. With the help of Alec and me, and none from his master who just stood chuckling and apparently enjoying the chaos, he soon rounded up the broadcast flock and had chivvied them into the field. I now had to weigh up the possibilities for transferring twenty rebellious sheep across the shore and into the launch and, at the same time, adequately containing the remainder in the field. I decided to leave the old dog on guard and the three of us would each have to carry an animal to the boat and as quickly as possible. We had lost nearly half-an-hour.

Imagine eighty pounds of squirming ewe, immensely strong and quite determined to escape whatever fate was in store for her. Alec picked up the first with accustomed ease, catching the creature by all four legs and lifting her over his shoulders before she realised what was happening. No problem. He stood waiting for me. I lunged for the nearest who, forewarned by her sister, was ready for me. I tried to grab her by the scruff and her silly little tail, but she wriggled away and ran for a corner of the field. I trapped her there and this time tried for legs gripped together, fore and aft. She kicked, struggled, protested loudly and all but broke free, but eventually she was precariously strung around my neck. Securely held there, she gave up the struggle and I joined an Alec who was splitting his sides. We set off for the Mary Ann, each with a docile load on board, and a word to John to get a move on and follow.

Alec strode across the shore. I laboured along behind, noting madam heavier by the minute and wondering how we could possibly manage to complete the job in time. At last I joined him beside the launch and we dropped our indignant ladies into the bottom.

'My god, you're fit,' I gasped, admiringly.

'He certainly isn't!' was Alec's grim comment, looking over my shoulder.

I turned to see how John was faring and it was a depressing sight. The old man was still some distance away, stepping with exaggerated care over the shore, stumbling, recovering, nearly tripping again. His eyes never left the ground, as if he was uncertain of the way and hardly knew where he was. As he came closer, we could hear him swearing long and loud. He literally staggered the last few steps to join us and then stood swaying uncertainly. The man was obviously in a bad way, his breath coming in painful gasps, his face a fiery red and the sweat pouring down it. I wondered how long he'd been drinking or, once again, was he ill?

'We'll have to carry the bloody lot,' swore Alec.

"Fraid so,' I agreed and then, thinking of the tide again, turned to see how the launch was positioned. 'Bloody hell!'

Our two brave ewes had recognised and seized the moment when our backs were turned. Both were clambering over the side. There were two resounding splashes and then our redoubtable ladies were in the water, paddling energetically for the shore. We grabbed each by its neck and tail and, with strength from somewhere, carried them back to the boat and unceremoniously dumped them into the bottom. Then Alec collected John's animal.

'He could mind the beasts while we go for the others,' I suggested on his return.

'Hm.' He sounded doubtful.

'Keep him out of mischief,' I added thinking of the whisky bottle back in the cottage.

All this while John had been standing on the shore watching the proceedings but taking no part in them. I hoped he was pulling himself together. When we rejoined him, I thought I caught a sardonic gleam in his eyes – satisfaction, maybe, that we did his hard work for him. Certainly, relief spread over his face when I told him of our plan.

'Okay, Don,' he agreed, with alacrity. 'I'll watch the buggers.'

Careless of a wetting, he immediately started wading in and in a minute was trying to climb into the launch. We had to help him over the side and then he sank, thankfully, on to the seat in the stern. As we turned away, he was already drawing his pipe from his pocket.

'We'll have to work fast, John,' I shouted, from the shore. 'Watch that rope. Give her more as the tide takes her down.'

'Don't worry,' he had the cheek to respond. 'We'll be fine.'

It took another forty minutes before we had carried seventeen more ewes across the shore and placed them into the launch. Alec did nine journeys to my eight and we were both shattered. I had kept an eye on John but he seemed all right, happily watching his charges and whacking the odd miscreant, intent on escape, with his blackthorn stick. However, when the time came to cast off, he rose unsteadily to his feet, was obviously shaky, and had to be helped to dry land. I wondered if the imminent departure of his sheep was, after all, going to be too much for the old man. But when he turned to gruffly thank me, all was made plain. A strong smell of whisky was on his breath and, with that, I noticed the tell-tale bulge in his pocket. For a moment I thought to make him hand it over, but what was the point? There was certainly more of the same in the cottage.

'We're away now, John,' I said. 'We'll be about an hour. Keep off that stuff and make sure the rest of the animals are safe.'

He grinned unashamedly then tottered off across the shore, weaving an uncertain path towards his home. Old Ben could be seen at the gate to the field, patiently watching his flock.

The journey over the loch was uneventful. There was no wind and practically no swell. The sheep settled quickly – packed like sardines in a tin they had little choice. Alec and I chatted in desultory fashion about the problems of old chaps who lived in remote places – they nearly always became a law unto themselves in all matters including their dealings with Authority. Mostly we sat in companionable silence eating our 'pieces', enjoying the warm sunshine and a much-needed rest.

A large cattle float awaited us on the other side. Expertly backed to the concrete ramp which ran down into the loch, its doors were gaping wide, its wooden ramp in position ready to receive the first load. As we tied up, two men sauntered down to greet us: 'fine day,' one said cheerfully, 'nice lot,' said the other, looking critically down at the sheep. Introductions followed. The owner and driver of the lorry, a cheerful, rotund sort of person, was Bob MacNab and the shepherd, tall, swarthy and lean, who worked for the new owner of the sheep, was Davy MacPherson. It seemed that they, too, thought whisky was necessary to the occasion, perhaps to celebrate a good day out. The fumes blasted our way and I reckoned both were pretty 'merry'. I hoped we were not going to have trouble and caught Alec's eye. He shrugged his shoulders and grinned.

I need not have worried. The animals were transferred from boat to lorry without incident. We lifted them, one at a time, on to the ramp and the men drove them straight into the vehicle. The whole operation took only fifteen minutes and we had made up some time. Just as we were about to cast off for the return journey, the shepherd made a suggestion that was to have unfortunate consequences.

'I'll come, too,' he offered. 'Get the job done quicker. I'd like to have a crack with John.'

For a brief moment I wondered whether the man could really be of much use – it would mean an extra body in the launch, for one thing, and would John welcome a complete stranger on this saddest of days? But there was no time to hang about and perhaps extra help, with the old man out of action, would be useful. I agreed and we set off at once. On the way over, we told MacPherson something of John's story and how the old chap was upset at having to part with his sheep. He was full of understanding: what a sad business, he would be glad to have a word with the old fellow.

When we reached Mark, all was quiet. The place looked idyllic in the autumn sunshine. Old Ben was still patiently in place, guarding the sheep behind him in the field, but, once again, there was no sign of the old man. Suddenly, I was uneasy. He should be out there with Ben, keeping an eye on the flock. Where was the old so-and-so? I asked Davy MacPherson to remain in the boat while we went to warn John of his arrival and to discover what was going on. We found John sitting on a bench against the back wall of the cottage. His blackthorn was between his knees, ready for action should the need arise. His pipe was in his mouth and the inevitable bottle stood on the seat beside him. I was furious.

'What is the meaning of this, John?'

'Sat down for a spell,' he explained, his voice thick. 'Must have dropped off.'

He was just sober enough to remember why we were there and to realise how angry I was.

'Sorry, Don,' he mumbled apologetically, struggling to his feet. 'I'll be all right in a minute.'

There was no point in losing my temper.

'Look John,' I said, trying to be patient, 'we've brought the shepherd over to meet you. Get your head in the sink. Get sobered up.'

'I'm not drunk,' he exclaimed with dignity, then lurched off round the corner to the cottage door.

Alec began to laugh. 'Silly old devil!' he spluttered.

'You wait,' I replied, grimly. 'He'll bugger things up, yet.'

It was at this moment that the shepherd appeared. He had thought us a long time coming and something might be wrong. Unfortunately, he had brought his dog with him and Ben was not used to strange dogs. John's dog rose at once, stiff with annoyance and the fur on his back bristling with suspicion. He bared his teeth and began to growl. MacPherson's dog wagged his tail, sank submissively to the ground behind his owner, and remained still. But the old dog was not satisfied. Snarling wickedly, growling fiercely, he advanced still nearer. The intruder stood again. Then both dogs began to circle around each other, hackles raised and looking for a fight. The shepherd called his dog. I called Ben. Too late. The tension broke and the two were at each other's throats.

We were trying to pull them apart when John, with dripping face and towel in hand, appeared at the cottage door. His befuddled mind saw Ben in trouble and needing help. What better help could an old dog have than his mates at his side? He threw open the door of the shed. Out raced the rest of the collies, each spoiling for a fight. The shepherd whistled and shouted. John shouted and swore. To no avail. Bodies, black, white and tan, became inextricably entangled in a snarling, snapping, yelping amalgam of canine ferocity. It was total confusion and a glorious doggy battle.

'I'll fetch a rope from the boat,' shouted the shepherd above the cacophony – and thereby solved the problem.

He began running for the shore. His dog saw him at once. The instinct to follow, to not be left behind, was strong. He broke away and ran off to catch his master. That was that. The enemy was routed, the scrap over. We persuaded the young dogs back into their shed and Ben set off after the few sheep which had escaped in the melee. Soon he had them safely back in the field. The old dog looked rather pleased with himself.

More time had been wasted. It was necessary, now, to move fast. The shepherd had not returned – perhaps he was consoling his dog and looking for injuries. His help must be enlisted to carry the sheep or we would not make it. Once again, I would leave Old John minding the sheep in the *Mary Ann* and I hurried to the kitchen. I found him with his head in the sink.

'Right,' I said, briskly, noting the bottle empty on the table. 'I'll take you over to the boat to meet the shepherd. He's Davy MacPherson. Then you can watch the yowes for us.'

The old fellow came meekly enough, still pursuing a rather erratic course, but sober enough to be guarding the sheep. Unfortunately, he was also sober enough to enjoy a chat with the shepherd. It turned out that MacPherson's uncle had once worked for John in the days when he had his own croft. They had much in common and took to each other at once. In no time at all they were nattering away remembering old times. I could hardly break up this pleasant exchange right away so, my mind on the falling tide, I hastened away to collect the first of the ewes, suggesting the shepherd might follow shortly.

'I'll be right up,' he stated, cheerfully. 'We'll just have a wee crack first.'

Then began a repeat of the first exhausting trip. As before, we left Ben watching the sheep in the field and then, one by one, yanked rebellious animals over our shoulders and marched with them down to the launch. Each time we arrived, we found the two friends deep in reminiscence and sharing the shepherd's flask between them, the dog sitting watchful in the stern. The *Mary Ann* rocked gently to the movement of her cargo and, from time to time, John's blackthorn thumped down on a woolly back.

'He's had enough of that,' I suggested to the shepherd each time we were there.

'Oh, aye?' he always replied cheerfully. 'It's just a wee nip.'

By now MacPherson was also sufficiently under the weather to make him virtually useless with the sheep. I gave up hope of an extra hand and comforted myself with the thought that the animals must be relatively secure with two men and a dog to mind them. John still seemed aware that he must release enough rope to keep the launch afloat. When, at long last, the final animal was on board, we 'heaved' Old John over the side and frogmarched him back to his cottage. We laid him out on the couch as best we could, then left him to sleep off his 'celebrations'.

The end was in sight. One more journey across the loch and we'd be done. Again we were lucky. Though there was not a breath of wind there was no mist and it was calm. The sun was pleasantly warm for an October day and once clear of the bay at Mark we let ourselves relax. Alec perched in the bows puffing contentedly at his pipe, from time to time flexing weary arms and shoulders, and I sat with the tiller beneath my arm, aching too, but thankful the day was almost over. From time to time, MacPherson's head nodded on to his chest, jerked back up for a moment then flopped once more. At last, with a shamefaced grin my way, he stretched out on the bench and fell sound asleep. Fat lot of use he was.

I suppose I didn't realise how tired I was, though I don't think I actually nodded off. Suddenly, there was a warning shout from Alec. 'Look out, Don!'

I shuddered back to reality and, horrified, saw an enormous piece of driftwood immediately ahead of us. Huge and unavoidable, it was only yards away.

'Bloody hell!'

I swung the tiller over and opened the throttle as far as it would go. The *Mary Ann* responded slowly. Too slowly. There was another yell from Alec, a horrible crunch and then the sound of splintering timber. The launch keeled over to starboard and began to take on water. The sheep, unable to stand, tumbled helplessly towards the side and over into the water. The shepherd, suddenly awake and not entirely 'with it', rolled helplessly after his flock. Alec, just in time and hanging on for dear life with one hand, managed to grab him with the other. The dog came slithering past. I seized him by his collar.

It looked like we were in big trouble, but the good old *Mary Ann*, bereft of her heavy load, shuddered and then righted herself. Dead slow, I was able to steer full circle back to the scene of the disaster. I cut the engine.

'Christ,' I said. 'What do we do now?'

MacPherson, all at once sober and his shepherdly instincts thoroughly aroused, took action.

'The poor buggers,' he shouted, and we watched in disbelief as the madman threw off his jacket and kicked off his boots. Before we knew it, he was on the bench and poised to jump.

Alec dived and caught him by the ankles just in time. 'Sit down, you idiot,' he yelled, as he hauled him back. 'What good will that do?'

'They'll drown,' the shepherd wailed, uttering the obvious, then crouched in the bows anxiously counting heads.

So far as I could tell the launch had not been badly damaged. She did not appear to be taking on water and though plenty had come in over the side, there were two balers on board.

'Any suggestions?' I asked Alec, as we worked away. 'We can surely pick up some of them.'

'Wait a little,' was the surprising reply. 'They won't drown yet. Let's see what they do.'

With just enough way on her, I held the *Mary Ann* against the ebbing tide and anxiously watched the gallant armada. Each animal frantically treading water; each with her head held bravely above the ripples of our wash, they paddled aimlessly about not knowing what had happened to them or what to do. How long before those woolly coats, at present buoyant with air, became sodden? Drowning, then, would be only seconds away.

A strange thing happened. As with all herd animals, there was a natural leader. I picked out the matriarch of the group by the markings on her head, a redoubtable old lady who should long ago have been turned into mutton. Suddenly, she knew what she must do. Feet kicking urgently, she manoeuvred round to face the distant Mark then, with unfailing instinct began to head for pastures old and familiar. Her sisters got the message, too. One by one, they lined up in her wake and began paddling bravely to catch up.

'Jesus Christ!' exclaimed Alec.

'They'll never make it,' I lamented. 'We're nearly halfway across.'

'The poor bloody things,' was all the shepherd came up with.

'Wait a little,' Alec counselled once more.

He was right. In only a few minutes, the old ewe hesitated. She stopped swimming, trod water, and was obviously ill at ease. Mark was too far away. She had suddenly realised it. With a purposeful air, she altered course, and in a smooth circling movement brought herself round to face the launch. That was the nearest 'dry land'! There was safety? She set off to join us and her faithful followers, one by one, turned to copy her example. Soon they were all swimming steadily our way and one by one, the bedraggled swimmers reached the launch. Never did two people least resent a soaking as Alec and I hauled the sopping bundles back on board. MacPherson busied himself comforting the poor souls, but his dog was more useful. He sat bolt upright beside me, alert for any trouble.

We reached the far side of the loch with no further adventures and the sheep, by now, too bemused to be rebellious, were quickly off-loaded on to the ramp and into the float.

'You got 'em all!' the driver exclaimed, admiringly. 'D'you think they'll be all right?'

'We'll throw them into the loch at home,' MacPherson said. 'That'll get rid of the salt.'

His charges safe, the shepherd seemed completely restored. We left him helping Bob MacNab check they were all securely locked in, then quickly cast off and set course for Mark. The last we heard of Davy MacPherson, shepherd extraordinary, was his stentorian voice raised to the heavens: 'Eternal Father, strong to save'. I had my binoculars. They revealed the lorry still there, a cab window open, and his arm waving a farewell. In his hand was a bottle. Presumably he was giving heartfelt thanks for a safe delivery from drowning. Whose health he was drinking, we could only guess!

'We'll need to see Old John again,' I said, when we had stopped laughing and were on our way. 'Better check he's okay.'

'The old chap ran true to form, right to the end,' chuckled Alec.

'I know. I hope he'll be all right. He really cared for those animals.'

Back, once more, at the cottage we found the old man sitting at the table in the kitchen, a pot of tea, not a whisky bottle, to hand. He poured us each a cup and we sat there sipping the hot drink gratefully, assuring him the sheep were safe, passing the odd commonplace, but all the time conscious of unspoken thought heavy with reproach. I glanced covertly at the worn old face, the shaking hands, the shabby clothes, and then out of the window to where there were no sheep grazing contentedly where they should not be. How would he fare?

'You'll be better off without them,' I remarked, at last. 'They were getting too much for you.'

He looked at me straight, honest blue eyes sardonic, vaguely troubled. 'Aye,' he said. 'Thanks.' Then turned away to his tea.

It was time to go. Pointless to prolong a difficult moment. We thanked him for our refreshment, promised to come again soon, then left him to come to terms with a new way of life. As the *Mary Ann* puttered slowly out of the bay we looked back. The afternoon light was deepening into darker evening. Mist, damp and depressing, was forming over the forest. The old cottage stood lonely and alone, for the moment no smoke lifting from the chimney. Ben lay apparently asleep on the garden path. The younger dogs raced around, working off high spirits. All were collie dogs out of a job and none in prospect. Old John stood motionless in the doorway, he, too, a shepherd without any sheep. He did not raise his stick in farewell.

I shivered. 'Let's get out of here,' I said.

ten: the last of the waterfalls

TRAGEDY struck twice in Ardgartan. One morning, just before I left the house for work, Jimmie phoned. It was about two months after we had rid Old John of his sheep.

'They're worried over at Finnart,' he said. 'Old John has not been seen for several days and the dogs are making the hell of a row. I think you'd better go down.'

'Doesn't sound too good,' I replied, instantly conscious of a sad old figure standing in the doorway of his cottage. 'I'll go this morning and take Alec.'

Alec had been out on the hill since dawn, stalking stags, but as soon as he came in and had had a cup of tea he was ready to go.

'Any luck, this morning?' I asked, as we set sail for Mark.

'Aye, I got a good royal above the Douglas Wood. I'll need to bring him in as soon as we're back. D'you think Old John's on a binge again?'

'Don't know. I wish we could get him to live nearer the village.'

'Aye, he's too old to be on his own.'

We travelled in silence, each intent on keeping as warm as possible on a grey December morning and neither of us anticipating anything worse than the old man with another big hangover. I had paid him a short visit at the beginning of November to see how he was coping without his sheep, and Alec had already taken supplies down to keep him going for a month. The old fellow had seemed reasonably cheerful and there had been no indication that all was not well. I had even planned to persuade him up to Arrochar for the New Year festivities.

We chugged slowly into the little bay and dropped anchor. Cloud hung thick and low over the forest and obscured the ridge above. Rain had been heavy during the night and Old John's burn thundered through the larch wood, white water tumbling down on its way to the sea. Everything was dripping. The cottage looked derelict and desolate, its door closed, the curtains drawn, and yes, there was no smoke rising from the chimney to indicate all well within. Something else was missing, too. Much as I disliked sheep for the damage they did, John's flock had lent a certain ambience to the place, a croft viable as a way of life and a home, battered by the elements, but the centre of all activity there. An eerie silence hung over the whole area. No sign of the man who should have been at work in the plantation. No high-spirited collies running around. I began to be uneasy.

'He's sleeping it off,' said Alec, but he did not sound too confident.

'He's been a long time doing it, then,' I retorted, remembering the message from Finnart.

As we walked up the path through the neglected garden, bedlam broke out in the shed. The dogs had heard us at last. But no dilapidated old figure opened the cottage door when we knocked.

'Better see what's going on before we let them out,' I suggested and we hurried into the kitchen.

It was dark, the fire was certainly out and in spite of the confusion of clothing, boots, an old newspaper on a chair, and all the rest of the usual paraphernalia, there was that indefinable feeling of a place deserted, no longer lived in. Particularly alarming was no jumbled collection of tinned food and packets of this and that on the old deal table, nor in the cupboard John called his larder.

'What the hell has he done with it all?' asked Alec, amazed. 'I brought him enough for at least a month.'

'Probably fed it to the dogs!' I declared flippantly, but thinking it more than likely. It would be less trouble than making them porridge.

I ran upstairs to make sure the man was not ill. I checked both rooms, though neither looked used – he probably made do with the couch in the warm kitchen below. Not there, anyway. Meantime, Alec had looked in the closet in the garden, John's 'bog'. No luck.

'You know something?' he said, when we met up in the kitchen again. 'I think he may have set off for Finnart to see the MacWilliams. Ran out of food. Ran out of whisky.'

'You're right,' I exclaimed. 'I should have noticed. His boat's not on the shore, is it?'

We looked at each other in dismay. The family on the other side of the loch had not seen him in the plantation, or about the house, for several days; the dogs, shut in the shed, had been raising hell; a binge, if he had been indulging in one, would normally have been slept off on the couch, and he certainly was not there. Unless he was somewhere behind a wall, bush or rock, still drinking himself into oblivion with an endless supply of whisky, or the missing dinghy had been carried away by the tide and not rowed away somewhere by the man. It was all beginning to look ominous. We had to get searching at once.

'The dogs'll find him,' I suggested.

'Wonder when they were last fed and watered?' queried Alec. 'They may not be in a state to find anyone.'

We hurried to the shed with considerable misgiving. All was quiet and I hoped it was because the animals had recognised our voices and not that they were dead or dying. As we neared the door, one or two of them began yapping again and I thought I recognised the deep bass of old Ben. Our welcome was frantic, but brief. The dogs knew what was needed most of all. They slid past our legs, ran, trotted, walked, and staggered to the burn and then drank long and greedily. Poor things. They were thin, filthy and starving but, thank goodness, none had died. The stench coming from the shed, was appalling. We took one look inside then closed the door firmly. That was for later.

Those dogs realised something was wrong. Somewhat revived by their drinks, they returned with wagging tails and collie 'smiles' to welcome us again, but immediately turned away to investigate scents around the cottage and in the garden. I am sure they were looking for Old John. Then, a strange thing happened. We set off to walk the shore and instead of running ahead, they drew close in to our heels and trotted soberly alongside. We tried orders to 'seek', but they were reluctant to move more than a few yards away. It was as if they sensed disaster.

For form's sake, using binoculars, we searched the shores on the other side of the loch as best we could. There was nothing we could see. In any case, it was unlikely, at that distance, we would spot a small boat washed up on the rocks, and the MacWilliams had already looked in the bays and inlets either side of Finnart.

'The wind would bring him to this side,' counselled Alec, as we continued along. 'Come on pups, find John.'

The dogs made desultory attempts to pick up scent, but with the tides washing in and out, it would have long vanished from the rocks and pebbles of the shore. We walked up to the track which led from cottage to plantation, and they sniffed around but showed no great interest. We turned for the shore again and they slunk along behind.

'I think we should go a mile or two in each direction,' I said, noting their reluctance with foreboding. 'After that, if we don't find him, we'll take the launch and work towards the top of the loch. If he was in the boat, wind and tide have probably taken him that way.'

We came to the rocks of the little headland on the far side of the bay. There was no obvious sign of the boat, or John, but we had to be sure. It was while I was making my way to a spot from which we could easily climb to the top that I noticed old Ben. He was standing on a rock near the summit, ears pricked, tail tentatively wagging, looking at something on the other side. He uttered an excited yelp, then disappeared.

What followed was a bizarre repeat of another drama enacted when I made my first journey on foot to Mark. On the way home I had scrambled over this same headland to an adventure with a wild bull caught between two rocks on the shore. Now, we climbed as fast as we could to the top and looked down to the bay on the other side. Ben was standing on his hind legs, forefeet planted on the side of John's boat, which was jammed between the same two rocks, his tail slowly wagging from side to side. He was looking at a crumpled figure in the bottom and every line of the animal's body was a mute enquiry. The bull had got away. Poor John had not.

We dashed down, the collies now finding strength enough to run ahead. They trotted up to old Ben and then all the dogs piled into the boat. But something was wrong. They sniffed anxiously at the body, then perhaps realizing death, slunk out again their tails between their legs. We hurried past and climbed over the side. A quick look at Old John's eyes, a hurried feeling for a pulse that did not exist, and even a forlorn listening to a heart that did not beat, confirmed our fears. The old man was, indeed, dead.

'Do you think it was his heart?' asked Alec.

'I should think so,' I replied slowly, for we could see no sign of injury except a graze where his head had probably struck the side as he fell.

I suddenly realised that Old John could only have been here, jammed between those two rocks, for at most eleven or twelve hours. Had he been there longer, we would have found the boat full of water for it was spring tides at the moment. So what had happened? How long had the old man drifted, unconscious with the tides, being carried out into the loch on the ebb and rolling back again on the flow? How long had he been dead? We were never to know.

'We can't leave him here,' I continued. 'The boat'll fill. Let's get him back to the cottage.'

The dinghy seemed undamaged. The tide was rising fast. We waited until there was enough water then, with both of us heaving, managed to free her. I jumped in and kept Ben with me. Alec would walk back with the other dogs, shut them up and then come to meet me on the shore close to the cottage. The journey did not take long, the water calm, the old boat sound. I ran her up as high up as possible on to the pebble shore and Alec was already there to meet me. We carried the old man into the cottage – no great weight in his wasted body – and laid him on the couch in the kitchen. We fed the dogs on some porridge and biscuits, all we could find, then encouraged them over the shore towards the *Mary Ann's* dinghy. Ben refused to go. I took him back to the house and left him lying on the floor beside Old John.

'We'll be back, shortly,' I said. 'I'll get him then.'

We sailed as fast as the *Mary Ann* would allow and I took her straight into the bay which was close to Jimmie's house. He was shattered. 'I should never have allowed him to stay on down there,' he kept saying and seemed incapable of pulling himself together and giving any orders.

In the end, I made the necessary arrangements. The local bobby and a doctor accompanied us once more to Mark and, there being no suspicious circumstances, we were able to lay John as decently as was possible in the bottom of the launch and take him up to Arrochar. Ben lay close beside the old man and it was a sombre, silent journey up the loch. It was a heart attack, as we had thought and the doctor decided the old man had been dead for at least three days. Death by natural causes was the official verdict. We gave him a grand funeral, for he had no relatives, and everyone who worked in the forest was there. The procession was long between church and graveyard, no hearse in those days, but shifts of six men, solemn and dignified, carrying the coffin and a piper leading the way. The 'wake' which followed lasted well into the night and would be remembered for many a year.

The circumstances of Old John's death stayed long in my mind and I mulled it over again and again in the coming weeks. In the end I managed to convince myself that, other than dying peacefully asleep in his bed, this had probably been the best way for the old man. A loner, who wished only to be alone, he had been too old to tolerate a move from Mark and a fresh start in the village. Jimmie Reid, however, was quite unable to come to terms with the disaster. He sank into one of his great depressions, blaming himself for the old man's death and could not be reasoned out of it. I reckoned he was drinking heavily and, certainly, he was never to be seen before midday or after. For me the affair was a lesson for the future. If ever I was in charge, no-one who worked in the forest, when over the age of sixty, would be allowed to live in such a remote situation.

After the funeral, life and routine in the forest returned to normal, but without the benefit of our head forester. Nothing unusual in that. It was a quieter period, anyway. Men and ponies had been transferred to another area where there was an intensive programme of thinning in process, so I was down to one squad of men and one of women. A new plantation was taking shape beyond Coillessan and in all but extreme conditions Archie was at work there on his tractor, ploughing the straight

lines over a rough hillside with his customary skill. The men, with their
ganger Andy Murray, were fencing. To the great relief of Katy, ex-Land
Girl and now proven pony girl, Mollie remained with us and she was able
to work her with the other ponies in one of the larch woods. The women's
squad, with Bill O'Connor, were at work brashing in a sitka plantation.
Each morning, the familiar crew of black, oil-skinned, sou'westered
'scarecrows' set off to what could only be called hard-labour in sometimes
very unpleasant conditions. The gales would blow, the rain come in sheets,
and sometimes when the temperature fell, it was a blizzard. When the
weather was really impossible, the men would be put to work in the shed,
cleaning and mending machinery and tools, and the girls sent home.

The weeks passed. The anniversary of my arrival at Ardgartan did not
receive special attention but I did note, almost with surprise, that I had not
thought about my health for many months. The effects of rheumatic fever
and the resulting problems with my heart seemed to have miraculously
vanished and were almost forgotten. I was fit, and I loved both my job and
my lonely old home in the Wood of the Waterfalls, where friendly ponies
whickered a welcome and the local wildlife made free of my garden. The
only continuing cause for concern was Jimmie Reid. He still showed little
interest in what was happening in the forest. We seldom met, seldom
spoke. Then towards the end of January, there was a second disaster in the
forest which finally drove him over the top and to a decision which also
had its consequences for me.

It was the old lorry. This was used for many purposes, transporting
men, machinery and tools, trees to be planted, and so on. Behind the
driver's cab was a small shelter and, behind that again, the 'well' into
which most of everything, men or materials, was loaded. This had hinged
sides, which could be opened outwards for loading and unloading, and it
was on these that in lieu of seats a few of the men would perch. It was a
dangerous practice but one which offered some degree of comfort on a
long, bumpy journey through the forest. The ancient vehicle would set off
in the morning, crammed full of squatting figures and those hanging on
for dear life to whatever was handy.

One morning, when the frost was thick on the ground, it left the yard
at Coillessan at the start of the working day, but within the hour had

returned with only the driver and Andy Murray on board. The ganger jumped down and ran to meet me. He looked distraught and my heart missed a beat.

'There's been an accident,' he panted.

'Take it easy, man. What's happened?'

'It's Wattie Reid. He fell off the side and hit his head. I think he's dead. For God's sake, come.'

'Where is this?'

'About halfway to Arrochar, at the side of the loch. There was ice on the road.'

I phoned for an ambulance, then we set off back to the scene of the accident. While we drove as fast as possible over the rutted road to where it joined the main road to Arrochar, Andy told me what had happened. Coming back from the village with his load of workers, the driver had stopped at a regular pick-up point where the two remaining members of the squad were waiting. In order to be out of the way of any traffic, he had pulled on to a concrete slipway, which led down into the loch, reversing in for an easier get-away on the icy surface. The two men had climbed on board and were moving to seat themselves on the side, when the driver, impatient to make up time, had let in the clutch too suddenly. There was a fearful jerk and a screech of skidding tyres. The lorry kangarooed forwards and Wattie Reid had fallen backwards and out. On a downward slope, the fall was longer, and harder. His head hit concrete with a tremendous whack and he lay very still. Andy told the men not to move the unconscious man, in case any injury might be made worse, but he felt, himself, that the man was already dead. No telephone booth handy, he had come as quickly as he could to report to me.

We turned on to the tarmacked road and soon drew up beside a forlorn group, all muffled up against the biting cold and stamping their feet to keep warm. Reid lay where he had fallen. I went through the motions of resuscitation but I knew there was no real chance. The poor chap was dead and I thought it likely he had died instantly. The ambulance arrived shortly afterwards, together with the local bobby who had hitched a lift. I left them to cope then ordered the lorry back to the Office. Jimmie Reid must be told at once.

In fact, Jimmie never recovered from this second blow. He was sitting at his desk reading a report when I told him the news and I thought he was going to faint. He sat there swaying slightly, staring incredulously at the foreman and myself, and for a long moment was silent. Then, having assured himself that I had called an ambulance and that the police were now in charge, he muttered a gruff dismissal. As we made our way out through the cottage door, we heard him diving for the little cupboard in the hallway where the whisky bottle was hidden. I did not look at Andy.

The inevitable enquiry by the police was held and no blame for the accident levelled at Jimmie. It had just been unfortunate. But, if it was possible, he became even more withdrawn and unapproachable and I reckoned he was blaming himself for both tragedies. In a way, I suppose, he was right. Old John should have been persuaded away from his lonely cottage and the men told not to sit on the sides of the lorry. One day, he sent for me. To my surprise, I found a cheerful boss seated in the scruffy little office, miraculously restored, wig in place, clothing spruce, and a pipe in his mouth. The reason became quickly clear. A decision had been made.

'I've something to tell you,' he said, quite briskly for him.

There was a pause during which he seemed to be searching for words already thought up, but suddenly forgotten. I wondered what was in store. Had he bad news to tell me?

'I've decided to retire,' he said, at last, the words coming in a rush. 'A forester called Duncan MacGuire is coming in a few weeks, but you'll be in charge meantime.'

'Are you sure?' I stammered, taken completely by surprise. 'You've years ahead of you, yet.'

'I want clear of all this.' He waved his arm in a vague sweep which included the familiar mess of papers, books and maps, and the forest outside only dimly seen through a dirty window. Then, true to form, he at once became deeply immersed in a document and I recognised dismissal. I received no further details of his successor nor further reasons for his departure, and enquiries, I knew, would receive only monosyllabic replies. On the bike ride back to Coillessan, I pondered this latest development and wondered what it would mean for me.

Paramount was dismay at the thought of a new Head Forester. I had just got used to this awkward one!

Jimmie Reid did not go out with a bang but he certainly went in a hurry. The decision made, he could not get away fast enough. There was no function at which a presentation and speeches could be made. Not even a party at his favourite pub in the village. But when the squads on pay day squeezed into his tiny office, all received a farewell handshake and a dram. The next morning, a furniture van arrived at his door and he, with his goods and chattels, disappeared from our lives. To my surprise I was quite sorry. In spite of his tiresome ways, I had learned to rub along fairly comfortably with a difficult man. I never forgot the famous telescope, though, and often imagined it following me around as I went about my business.

Our new forester would not be with us till the end of March, so for a blissful month I would be in charge. I have to admit I revelled in the thought. After all, to a greater or lesser degree, Jimmie had opted out long ago and in all but name I had been doing the job. First on my list was the office. As a young forester newly out of college, I had ridden my bike along the rough road from Coillessan to report to Jimmie Reid at the Forest Office. I had found a brand new building not in use and my first interview had taken place in that tiny room in his own house. Now, I spent a morning carting maps, papers and files to their new home, arranging for the telephone to be connected and a hooter to be installed in case a fire warning was one day needed – I hoped that it never would be but we were fast approaching that time of the year when the danger was high. At last, we had a proper centre from which the forest would be run but I continued to live in my battered old house at Coillessan and arranged for urgent calls to be diverted there when necessary.

The powers that be now rated the post important enough to have a van provided for the forester's use. Let it come soon, I thought, for spring was nearly here and I would need to be everywhere at once. Spring also meant wildlife waking from its winter sleep and the breeding seasons begun. I was determined to find time for watching the eagles, badgers, foxes and otters there had been no opportunity to watch before. And the work went smoothly enough. There were no alarms and no problems I

could not handle. The only minor hiccup was when one day Katy came
to see me in the Office. There she demurely told me that she was leaving
because she and Peter, Tommie the pony's handler, were planning to get
married. He had been moved to Carnach and she would be following
him there as soon as she could.

'Do you think Mollie could come, too?' she asked, ingenuously.
'Perhaps we could exchange with someone?'

'I'm sorry, Katy,' I replied firmly. 'I can't muck the teams about just
because you're getting married.'

The honest blue eyes regarded me with scorn and I really believe she
thought I was just being awkward. The funny thing was that she never
did leave, but that was her affair, not mine. Perhaps she loved Mollie
more than Peter.

As the weeks passed, I began more and more to regret the imminent
arrival of Jimmie's successor. I was managing fine on my own! At the end
of long days in the forest or out on the hill, I would dream of the perfect
forest Ardgartan could become – with my careful planning, of course.
There would be woods of pine, larch, fir and spruce, a nice mix on the
hill, which would be a delight to a forester's eye. This must surely be the
perfect career, I thought, trees and wildlife, their conservation and
welfare my concern.

One evening towards the end of the month, I was biking back to
Coillessan. I looked up to a beautiful larch wood where a pale green flush
proclaimed spring almost with us. I remembered someone reporting
having seen red squirrels there and thought I would check. It had been
a glorious day, sunshine and showers, and warm. I trod on a carpet of
needles and noted cones lying beneath some of the trees, each one
stripped of its seed in squirrel fashion. I found two squirrel 'tables', one
on a stump, the other on top of a moss-covered rock, and each with its
residue of abandoned cones. Three dreys, each built high in a tree, were
cradled safely in swaying branches. To crown all, as I sat down beneath
one for a breather, scolding came from above. A rust-red creature with
sharp eyes and bushy tail reminded me that this was its place, not mine.

I lingered to enjoy the last of the day and remembered an unopened
letter in my pocket. I fished it out and noted it was 'Official'. No alarm

bells rang, for a fair number arrived at the office. I opened it casually, all the while keeping an eye on the squirrel. But squirrel decided it was time to go home, clawed along its branch and popped into its drey and I remembered the missive in my hand.

It was brief and to the point. Two foresters, not one, were coming to Ardgartan at the beginning of April and after a suitable handing over period I would be transferred to Fearnish, a forest further east. A house was available.

It was a bitter blow. What had I done wrong? How had I offended? A phone call came the next day. It confirmed the transfer and explained all. The move was simply to give me a different experience and I could regard it as the first step to future promotion. It was little comfort at the time. I had come to love this area with its enchanting Wood of the Waterfalls, its beautiful old forests and its promise for the future. I had a workforce, too, with whom I got on well and had no problems of discipline or lack of co-operation. I had thought to be there for several years at least.

My road to Ardgartan had been long and devious, my sojourn there short. What lay ahead?

epilogue

SEVERAL forests lay ahead. Each one was different in age and scale and each represented a step forward in a varied experience. Changes in forestry practice were already taking place when I left my first forest and these continued over the years. With the need to increase the forest resource after the depredations of two world wars and, in due course, to meet the present day demand for timber products, the attitudes of professional foresters had to adjust to different policies. Silviculture died a slow death and the economics of 'tree farming' gradually took over.

The Sitka spruce arrived to stay. This is a tree from North America which will grow well and quickly in almost any soil conditions and in most extremes of climate. In addition, unlike other species, it will withstand the browsing of deer. Its potential was immediately recognised and there was a huge increase in its use. Unfortunately, it is always grown as a crop, the trees planted most often in rectangles with 'rides' imposing a regular pattern throughout, the edges of the forest as near to a straight line as is possible. It is grown to a certain height and girth, then harvested, as a clear fell, to expose bare and forlorn hillsides – until the young trees of the next rotation appear. Instead of true forest, vast areas of trees all the same height, colour and shape have become the norm. For this reason Sitka spruce is often maligned. Which is understandable but a pity. If allowed the space and time to grow to maturity, say 150 years, it can be a magnificent tree.

Mechanisation gradually took the place of manpower and had a profound effect on the people who worked in the old forests. Fewer and fewer of them were needed. The sweet rasp of the cross-cut saw (two men) was replaced by the harsh whine of the power saw (one man). 'Humphing', the manhandling of each log from the felled tree to a collecting point is a labour of the past and ponies (each with its handler) are now rarely used for hauling out the timber. Instead, huge mechanical monsters (one man) clear swathes into the forest felling the trees, stripping the branches from their stems, cutting the stems into lengths, and finally, transporting them to the forest road to await the timber lorry. The lorry driver, by means of a crane on his vehicle (one man instead of several and sometimes, a woman), then completes his load and takes it to the sawmill or some other destination. Even the fencing of a new plantation can largely be done mechanically, though it is still often undertaken manually. But what used to be the work of many men is now often completed by one and his machine.

There is one task, however, that is still be carried out by human beings. It is the planting of the trees. No-one, as yet, has invented a machine that will do it better or more economically and the sheer hard labour involved has not really changed. Doubtless this problem is being worked on and one day soon we shall see robots, linked to a computer, trundling across the hillsides, up and down the rows, meticulously inserting the transplant trees at exactly the correct intervals and in the shortest possible time.

With mechanisation and the requirement for a much smaller work-force began the demise of the forest 'villages'. These now tend to be home to commuting folk who like to live in the country but work in the city, or the weekend retreats of those with a second home. Gone are the expert woodsmen of the forest. Gone, too, are the cheerful squads of women who did a valiant job with few grumbles and not inconsiderable skill. There are still people in the forests who fence and plant, but they are generally contract labour from outwith the area. Only very few now live in rural communities.

The modern forests are, and have always been, regarded as blots on the landscape. It should be remembered, however, that these forests help to cover degraded hillsides almost devoid of the trees that once were there. Once upon a time, pine, birch, hazel and others covered the land to a greater or lesser extent, protecting the soil and offering habitat to a wide diversity of wildlife, some of which is now lost to us. Over the centuries, partly out of greed, partly out of thoughtlessness, man has been responsible for this unhappy state of affairs. The tree resource was over exploited, for the most part without proper management. Vast flocks of sheep were introduced in the 18th century, and onwards, and these overgrazed the vegetation and prevented the natural regeneration of trees. In the absence of its natural predator the wolf, and in the interests of the sporting estates, red deer numbers were allowed to increase to unmanageable proportions and the damage they have done to the environment is considerable. There are still far too many of both.

In the past, there were some enlightened estate owners who recognised that trees are necessary. From around the second half of the 18th century some of them planted excellent policy woods and ornamental avenues of different species; these often included exotics from other countries and continents. The woods of that time were easy on the eye and often a pleasure to wander in. Unhappily, when they came to be replanted, they became the much-criticised plantations of today.

Trees offer valuable habitat to both mammals and birds, but woods and forests evolve slowly and the creatures who live within them will have matched that pace, only being present when the habitat is suitable. While the forest is young, field voles are a major source of food for short-eared owls, hen harriers and kestrels. The sparrowhawk will replace them once the canopy has closed and instead will predate upon small birds such as tits and chaffinches. Mammals play an important role throughout the whole rotation. Pine marten numbers have greatly increased because of the available forest habitat. Red squirrels use larch and pine, in particular, but also build their dreys in both Norway and Sitka spruce, making use of their cones as a food resource. Red and roe deer thrive in forest and woodland and the former, so often obliged to live in the harsh conditions of exposed moorland, is a bigger and better beast because of the shelter provided by the trees. Tawny, barn and long-eared owls, goshawk and buzzard are there as the forest grows older.

In recent years, the voice of public opinion has been ever more vociferously raised in regard to modern forestry and perhaps there has been unease in the mind of the true forester as well. Trees are necessary. They do need to be grown, both to conserve and nurture the soil and to meet the insatiable demand by the human race for timber. But monoculture is neither good for nature nor congenial to the eye of the human beholder. In common with many a forester of today, I see no need for those unnatural blocks of Sitka spruce upon the hillsides.

Forests of mixed species can be made to be viable economically and trees can be planted to suit the landscape and the contours of the hillsides. Rather than unsightly clear felling and replanting, a continuous tree cover policy, with natural regeneration, should be the norm. Broad-leaved species must be planted, too, and these are especially suitable at lower elevations and far up the hillsides in the gullies. Existing remnants of natural forest should also be extended, but here there is a problem. A bare hillside devoid of 'mother trees' does not allow regeneration by natural means. Planting is the only alternative. This should not be a complete tree cover but a percentage. Not more than fifty percent. The remainder should be allowed to regenerate later, or even to remain without trees. Provided sheep and deer numbers are reduced, there is ample space in the Highlands for both a commercial and an amenity for-est landscape, the latter planted with indigenous species.

That is how I began to see it, all those years ago. Not as a result of any scientific research, but instinctively, as Old Dougal the woodsman might have done. Unfortunately, trees take a long time to grow and the mistakes of the past cannot vanish overnight. But progress is being made and the future begins to look much brighter. Perhaps the long ago dreams of a young forester can still be realised.

afterword

IN the years that have passed between Don MacCaskill's appointment as a young forester, his retirement from the Forestry Commission, and until he died, his niche as a prominent nature conservationist, there has been a magnitude of change in the forestry industry, in both Scotland and the rest of the UK, which few could have foreseen. These changes have affected the landscape, the wildlife and those living in rural areas and have left us with good news and bad.

The last ten or so years have brought to fruition the ideals and hopes of many wildlife conservationists. Of course, no one person can claim responsibility for these changes, but as one who had the privilege of working with Don MacCaskill during the 'dark ages' of forestry and nature conservation, I do wonder just how much difference this man made to the speed and nature of this changing scene. I certainly believe that his wisdom, considered by many at the time to be of the 'left-wing, open-toed-sandals' variety greatly influenced people working within the forestry industry through the forthcoming years.

For example, his cunning avoidance of implementing the proposals to kill large areas of oakwoods on Loch Awe side in the 1960s was nothing short of disobeying orders, but it was done without a fight, through a quiet but deliberate refusal to get on with it. This was typical of the man and it undoubtedly accomplished so much more than a confrontational refusal would have done at that time. Today's legacy of these beautiful oakwoods is certainly in part a tribute to his commitment to doing what was clearly the right thing to do. It is interesting that the western oakwoods, referred to as 'scrub' at that time, and at best maintained as shelter for sheep, and at worst cheaply disposed of by 'ringbarking' to facilitate replacement by commercial tree species, are now recognised in European legislation as an important habitat for conservation.

In the 1950s and 1960s, the views being expressed by some of the eminent nature conservationists such as Aldo Leopold, Paul Ehrlich, Raymond Dasman and Frank Fraser Darling were considered radical and extreme by many, and were certainly not heeded by governments. Fraser Darling was a broad minded exponent of 'Sustainable Development', long before the term was coined by present governments, and his Reith Lectures of 1969, entitled 'Wilderness and Plenty', are as valid today as they were then. He considered that the Cuthbertson plough (which was being used at the time to reforest large areas of the Scottish Uplands), and the Sitka Spruce, provided a major benefit in reversing centuries of soil erosion and degradation by restoring tree cover. The type of forests being created seem to have been of less importance to him than the huge potential benefits associated with reforestation.

Perhaps he had envisaged that forest monocultures grown over short rotations were simply a necessary step towards the establishment of more diverse and natural forests? If so, we might later reflect on whether he was right.

In 1992, at the Earth Summit in Rio de Janeiro, leaders of over 150 countries met to agree to a more sustainable future for our planet. The Earth Summit was a major milestone along the pathway to encouraging governments and people to think and act in ways which do not erode environmental values and help us progress toward a better quality of life for everyone. The United Kingdom signed up to it. While many people in the UK may not be aware of the Earth Summit, those who are feel that little has changed since 1992. However, it does take time to develop ways of turning these ideals into local reality, and many people and organisations have been working very hard since 1992 to ensure that things really do change.

In 1993, European countries signed up to the 'Helsinki Resolutions', a series of principles for the protection and conservation of European forests. In addition, the Species and Habitats Directives encompassed within the Natura 2000 Programme have identified important European species and habitats which are in need of conservation management. In Scotland, these include capercaillie and red squirrels, native pine and oak woods, and work is well advanced to further their conservation.

1994 saw the UK Biodiversity Action Plan, produced to provide a strategy for delivering the promises made at the Earth Summit. Throughout Scotland, local communities are drafting Local Biodiversity Action Plans to identify needs and actions at a local level. We have indeed come a long way since the subversive protection of the Loch Awe oakwoods in the 1960s!

Don MacCaskill's working life encompassed the rise and fall of the forest community, from the enticement of people from urban unemployment to the remote forest villages to support the large afforestation schemes of the 1960s and 1970s, to the redundancy of local skills fuelled by the demand for greater efficiencies through more machinery and the employment of contractors from afar. The total employment in downstream manufacturing may be similar or even higher than before, but that does not help keep people in the small Highland villages and clachans. This is bad news and it is hard to find a solution to the depopulation of remote communities.

We should remember that these declines, though often reported as simple statistics, represent a loss in terms of the disappearance of a countryside culture and way of life. Will the villages being increasingly occupied by commuters and refugees from a poorer quality of life in suburbia ever generate replacements for Bill O'Connor and 'Old Dougal'?

In 1969, Frank Fraser Darling wrote

The near landscape is valuable and loveable because of its nearness, not something to be shrugged off: it is where children are reared and what they take away in their minds to their long future. What ground could be more hallowed?

Is there some reflection in these words of the impressions which that young boy from Kilmartin took away with him on his life's journey as a forester?

Where next?

Forestry is a thriving and important industry which is supporting a valuable wildlife and many jobs. The United Kingdom, largely dependent upon timber grown in Scotland, is now producing more timber than Norway. Society is demanding more from the land now than ever before, and we must be careful how we plan its future use.

In Scotland there remain enormous areas where the impact of

grazing sheep and deer are preventing native tree regeneration on land which is highly suited to woodland development. This is not to suggest that we should be striving to create some pristine wilderness of native forest reminiscent of a post-glacial or mediaeval age, but that we should build upon the desirable landscape features, land uses and future needs to develop a model on which we might build future landscapes.

We must safeguard the artefacts which express our cultural heritage such as the ancient stone circles, burial chambers and settlements. We must maintain open spaces such as the heather and grass moors which, though maintained by human intervention, have become part of the natural heritage and landscape of Scotland and which allow the interpretation of our unique landforms and glacial features such as the 'parallel roads' of Glen Roy. We must also maintain, and if possible enhance, the diversity of wildlife, the biodiversity which is so fundamental to the quality of the countryside. Finally, we must consider how forestry might help to reverse the depopulation of rural Scotland.

The interpretation of UK forest policy from the Earth Summit and Helsinki Guidelines is explicit in its commitment to people and communities:

Forest resources and forest lands should be sustainably managed to meet the social, economic, ecological, cultural and spiritual needs of present and future generations.

Our vision must therefore include a much greater involvement of local people in forestry, from the education of our children, the leisure and recreation of families and an economy which is based on a much wider diversity of forest products, including local markets for wood products, the sale of fruit and game and eco-tourism and recreation.

THE FUTURE OF SCOTLAND'S FORESTS

The great momentum which has been generated thus far by individuals and organisations such as the Forestry Commission to move on with an even greater determination to ensure that forests provide jobs, wildlife, pleasing landscapes and recreational opportunities must be sustained. Perhaps most importantly, we need to try to rebuild the lost 'forest culture' and increase our awareness of working with the forces of nature, an ethic which remains embedded in the psyche of Scandinavians, Poles, Hungarians, Czechs and Slovaks. The way to do this must be by ensuring that children at the earliest possible age are encouraged to interact with the natural world, learning to respect wildlife and natural places as things which will contribute to their own well-being. This can be achieved in urban as well as rural environments if the will is present.

While we are considering future generations, how can we know what future generations will want? This is made all the more difficult when we are planning, restoring and creating forests and landscapes which will only become fully developed over huge timescales. Perhaps the best we should attempt to do is to ensure that by conserving all our current ecosystems and habitats and enhancing biodiversity as effectively as possible, we are at least providing future generations with a greater range of options than is the case today.

We should perhaps provide much bigger areas of forests in the landscape. Some of these forests will be so large as to encompass villages and townships, similar to the Forest of Dean in Gloucestershire and to many forest villages in Europe. The forest will also encompass large non-wooded areas, forming a rich mosaic with woodlands interconnected by riparian corridors following the burns and rivers and allowing wildlife to move freely between woodland patches. Open ground species will not be isolated by the woodlands and a network of open habitats will be in juxtaposition to the woodlands, also providing many niches for species which thrive in edge or transitional environments.

Sheep will almost certainly play a lesser role in the economy of the Highlands and the benefits of managing deer at considerably reduced densities will be increasingly recognised. This will result in decreased dependence on the use of deer fences, and deer stalking will become more closely integrated with other land uses. Forestry and agriculture will become more closely associated, with farmers realising the value of farm grown timber for use on the farm. Niche markets for locally grown native and non-native timbers will increase and more portable milling facilities will be available. School curricula will include considerable opportunities for involvement in local industries. Art and poetry will once again focus on the local natural environment to establish a new age of forest folklore. Eco-tourism will expand to become a major component in the rural Scottish economy.

Although the balance of different management objectives will differ between woodlands, all commercial woodlands will have additional objectives. Native species will usually play an important role, especially in the larger woodland patches, but the current polarised distinction between native and non-native will disappear in all but the ancient semi-natural remnants. Instead, habitat networks of native woodlands and open habitats will intersect woodlands maintaining high native biodiversity and supporting the important commercial objectives of a continually increasing timber resource. Woodlands which have probably always been rather scarce, but which have more recently become rare will expand. These include the hazel woods which support such an enormous variety of rare lichens, and the aspen woods which support a unique diversity of insects which depend upon the dead and decaying wood of aspen.

Within this forest of native and non-native species, the result of some natural disturbances will be left to evolve naturally through colonization and succession into new woodland patches. The age and structure of the woodlands will become more varied, supporting a greater variety of wildlife. Minimal intervention areas will be identified in which natural biological processes will be the dominant forces of change. Adjacent to these, patches of woodland will be managed on long rotations, with some trees being left for between 70 and 100 years before harvesting. New markets will be developed which will welcome the big logs which will be produced. Much of the forests will continue to be occupied by non-native species and managed on short rotations to supply the chip board and pulp wood markets. Silvicultural methods will become much more diverse, employing a range of approaches from clear-cutting through small scale felling of groups of trees to the selection of individual large trees.

Throughout these woodlands, large amounts of standing and fallen dead and decaying wood will be one obvious change to the forests with which we are currently familiar. This important resource, so dominant in natural forests, will support a wealth of life from the most inconspicuous unicellular organisms, mosses, liverworts and lichens, insects and other invertebrates, to birds such as woodpeckers. Many of these will, in turn, aid the recycling process, eventually converting the wood to humus to become once again incorporated into the cycle of soil building and woodland development.

Finally, species which have become extinct due to loss of habitat and persecution might once again appear in our vision for the next century. Already sea eagles have been reintroduced and the reintroduction of the beaver is being seriously considered. While the reintroduction of these species is a logical and perhaps natural progression of our vision, it is important to consider whether the present and future land uses and habitats will be suitable. A forest with beaver, lynx, wolf and brown bear will certainly make for an interesting and diverse forest. This may not happen in our lifetime but, if we can continue the process of repairing the damaged forest ecosystems, who knows what the distant future might hold?

Philip R Radcliffe,
7th November 1999.

Dr Philip Ratcliffe began his career in forestry as a forest worker in Wales and worked as an assistant forester to Don MacCaskill in Argyll. In 1987, he became Head of the Forestry Commission's environment branch and in 1993, Head of the EC's environment branch where he was involved in developing policies aimed at enhancing the wildlife and conservation value of forests. This lead to his involvement with the UK Biodiversity Action Plan Steering Group and in post-Earth Summit developments in Europe and North America. In 1996 he left the Forestry Commission to establish a successful ecological consultancy and has since worked on a wide range of land issues. He is currently a member of the Scottish Natural Heritage West Areas Board and of the Deer Commission for Scotland. He has written over 50 publications.